Essentials of Metallurgical Thermodynamics

Prof. Dr. R.H. Tupkary
Former Prof. and Head,
Met. Engg. VRCE, NAGPUR - 11

Dear Readers,

To prevent the piracy, this book is secured with HIGH SECURITY HOLOGRAM on the front title cover. In case you don't find the hologram on the front cover title, please write us to at contact@khannabooks.com or whatsapp us at +91-99109 09320 and avail special gift voucher for yourself.

Specimen of Hologram on front Cover title:

Moreover, there is a SPECIAL DISCOUNT COUPON for you with EVERY HOLOGRAM.

How to avail this SPECIAL DISCOUNT:

Step 1: Scratch the hologram
Step 2: Under the scratch area, your "coupon code" is available
Step 3: Logon to www.khannabooks.com
Step 4: Use your "coupon code" in the shopping cart and get your copy at a special discount
Step 5: Enjoy your reading!

Price: ₹ 224.00

ISBN: 978-93-82609-032

Essentials of Metallurgical Thermodynamics *by*
Prof. R.H. Tupkary

First Edition: 2016
Reprint Edition: 2024

To view complete list of books, Please scan the QR Code:

KPH

Published by:
Khanna Book Publishing Co. (P) Ltd.
CIN: U22110DL1998PTC095547

Visit us at: **www.khannabooks.com**
Write us at: **contact@khannabooks.com**

Printed in India by: Vaibhav Products.

Dedicated To

All those scientists and technologists who are striving hard to raise the energy efficiencies of various engineering and technological processes in India to bring these up to the best standard in the world.

Acknowledgement

The author is indeed indebted to many in the preparation and publication of this book in a very short span of time. I am grateful to all my earlier colleagues and staff members of the Department of Metallurgical Engineering, VRCE, for their constant encouragement in writing this book.

The scientific tradition as exemplified by the following quartet

"वैज्ञानिकश्च्व कपिलः कणादः सुश्रुतस्तथा
चरको भास्कराचार्यो वराहमिहिरः सुधीः।
नागार्जुनो भरद्वाज आर्यभट्टो बसुर्बुधः
ध्येयो वेंकटरामश्च्च विज्ञा रामानुजादयः।। "

always inspires me to contribute my little might in regaining that scientific glory of our nation. May this piece of scientific work by my humble offering at the alter of our beloved motherland.

R.H. Tupkary

PREFACE

Thermodynamics is taught at under graduate level in the degree course and diploma course of metallurgical engineering all over India and yet no appropriate book is available for the beginners to understand this most abstract and conceptual subject. The available books are rather too elaborate and over exhaustive and an average student finds it difficult to separate out the required material for his purpose at his level. The non-availability of an appropriate book becomes an handicap in understanding this abstract subject. It may be one of the reasons why this subject is dreaded most by the students and the net result is that the subject, which ought to be understood properly for understanding other subject at higher levels in metallurgical curricula, is appreciated by the students only mechanically for the purpose of solving numericals alone.

Having taught the subject at the beginners level for over my entire teaching carrier difficulties of students became more and more apparent. The desire to write an introductory text book on thermodynamics for metallurgical engineering students became intense but availability of leisurely times was rather remote. This became available only while canvalescing after an accident in which both tibia and fibula were fractured.

I hope the students will like this new addition after the earlier two - steel making and iron making - equally well. Even if the students do not find it so useful as the earlier two books, the pleasure of preparing the manuscript of the book on thermodynamics always made me forget the pains of a fractured leg during the period of convalescence. And now when the book is being published, the unpleasant memories of the accident, if not fully, at least pontially, have been wiped out.

I take this as if desired providentially.

Date : 31st March, 2016

R.H. Tupkary

CONTENTS

Chapter 1

Introduction

Scope, Concepts and Terms

1.1 HISTORICAL

1.1.1 Energy in Historical Perspective

The stone-age man sustained himself by eating whatever was naturally available, *i.e.*, green fruits, roots, dead animals, etc. and perhaps resided in natural caves. He must have felt the warmth of the bright sun without knowing that it was due to the thermal energy associated with the sun rays. He must have accidentally discovered fire and felt the same warmth that he experienced in sun-rays. With the passage of time he gradually understood the need of cooking his food, albeit crudely, may be on fire lit by burning dried leaves and chips of trees. Slowly some crude implements made out of branches of trees or pieces of stones might have been developed by him and similarly he appreciated to build crude habitat made out of similar naturally occurring materials like bushes and trees.

He was moving around in search of food thereby doing some labour for sustaining himself out of natural instinct. In modern terms he was doing some work. The human energy was being used to do work. This is true of all beings of animate world, illustrated here by the example of human being. The food consumed by human being is the energy that is converted through biological and physiological processes into work. A dead man does not do such work because he does not consume food. In modern days the food consumed by a human being is assessed in terms of calories equivalent of food he consumes. A robust man consumes more food and, at least, is capable of doing more work.

As the human civilization progressed from stone-age into the bronze-age man appreciated and developed the art and science of extraction of copper and tin from their minerals, thereby producing alloy bronze and its conversion into useful implements for sustaining his civilization. The chemical energy of coal must have been used in this extraction operation without understanding it as is understood today. This is really the beginning of use of energy for changing the existing material forms, *i.e.*, copper and tin in minerals, into useful implements for better human life.

The same is the story for all materials that slowly came into use in the development of human civilization since the stone-age. The bronze-age progressed into iron-age to produce a much wider range and better quality materials for the use of mankind. This ultimately brought in the industrial revolution and progressed further into its modern form through the enormous engineering and technological activities.

All the modern manufacturing activities are nothing but changes brought about in the existing chemical, physical and mechanical forms into new forms. For example, iron ore is reduced by carbon to produce iron and then converted into steel. Steel is processed to produce varieties of forms and shapes. A tree is finally converted into a chair or a table. A bamboo is converted into a paper which is then made into a notebook. All these are changes deliberately brought into existing material forms for the benefit to mankind. Without these manufacturing activities the present day human civilization just could not exist.

In fact the history of human civilization is nothing but progressive increase in amount, quality and complexity of materials made available for the use of mankind. This was all possible because industrial revolution could obtain energy on an increasingly larger scale, hitherto unknown, for the benefit of mankind. No wonder the quality of human civilization is assessed in terms of per capita consumption of various materials, in particular the basic materials like steel, aluminium, copper, etc. or per capita consumption of energy like electrical energy.

Energy was available to the stone-age man as is available at least in the naturally occurring forms. He however could not use it because he could not harness it as is possible today. Industrial revolution is a **'revolution'** in terms of harnessing the available energy. The steam engine is the device that was invented first to convert essentially the chemical energy of coal, by first converting it into thermal energy, and then into mechanical energy and thereby started off the industrial revolution. Consequently the railway train, textile mills, tool manufacture and so on got off on hitherto unknown larger scale.

Soon chemical energy could be converted into electrical energy in thermal power plants and manufacturing activities carried out on increasingly larger scale using electrical power. Electrical energy could be generated by constructing dam, on rivers. The potential energy of water, so stored, was subsequently converted into electrical energy through hydel generators. Similarly chemical and electrical energy was interconverted as per need, through electrochemical cells to produce one from the other. This finally culminated into the development of almost infinite source of energy through conversion of mass into energy through nuclear reactors, which is nothing but a miniature form of celestial star, a perennial source of energy on earth.

If one looks at the history of human civilization from the point of view of energy utilization, few facts come out quite vividly.

1. The human energy is limited and hence the work that can be carried out manually is limited. It is estimated that a, human being can offer energy equivalent of about 1/8 horsepower.
2. External energy is required to obtain new materials and in their new forms.
3. Energy is required to produce materials in their proportion, quality, complexity, etc.
4. Use of more and better materials is responsible for the development of human civilization starting from the stone-age upto the modern super industrial civilization.
5. Industrial revolution is nothing but the development of means of harnessing energy for the benefit of mankind.

In energy terms therefore, the history of human civilization is nothing but development of means of harnessing energy in ever increasing proportion, from and in its varied forms, for the benefit of mankind.

The present day civilization sustains itself on its tremendous capacity to bring about changes in existing material forms. A change may occur as a natural one, for example, burning of coal, accompanied by evolution of energy. However most changes are to be brought about by supply of energy from outside. These are unnatural changes typically illustrated by the example of extraction of metals from its minerals. The necessity to evaluate the actual energy requirements for such changes and the progressive decrease in the energy consumption by way of the improvements in such activities has been one of the landmark development in human civilization.

1.1.2 Thermodynamics in Historical Perspective

The establishment of industrial revolution on a firm footing by the eighteenth century, as was the result of human ingenuity, was also the cause of further development of human ingenuity, quite naturally, compelled these **'revolutionaries'** - the men of science, to understand the scientific principles underlying the means adopted for harnessing the thermal energy readily available from burning of chemical fuels. The early approach was essentially related to the steam engine - a device that converted thermal energy into mechanical energy. This eventually crystalised into the study of this dynamism of thermal energy into what is now known as the **'science of thermodynamics'**. For quite sometimes even after the commencement of industrial revolution thermal energy was the only energy form available to be harnessed for useful industrial activities. This is the origin of science of thermodynamics.

With the progress of industrial revolution on the one hand and development of science as a whole on the other, led the scientists to discover several other forms of energies like electrical, electrochemical, magnetic, hydraulic, nuclear and so on. The harnessing of these energy forms involved the development of means of interconversion of these energies. The science of thermodynamics had to deal with each one of these interconversion of energies, as it appeared on the scientific horizon. The traditional name however continued to be used for all such studies although the interconversion at times did not at all involve thermal energy as such. Strictly speaking therefore the science of interconversion of energies need to be named anew as something like **'energydynamics'** to cover all forms of energies, but the same traditional name thermodynamics continues to be referred to the study of interconversion of energies.

The science of thermodynamics really developed over the last few hundred years but more so in the last two centuries. The formulation of Nerns't Heat Theorem in 1906 completed the formulation of **'classical thermodynamics'** ***i.e.*, the laws of thermodynamics and related mathematical** deductions. In should be remembered that the desire and the necessity to assess the efficiency of steam engines, then in use for quite sometime, more than anything else, led to the development of classical thermodynamics. Hence the related mathematical deductions are often illustrated with reference to a steam engine or in its idealised form as an ideal gas engine.

1.1.3 Scope of Metallurgical Thermodynamics

The science of thermodynamics here deals with only conventional forms of energies like, electrical, mechanical, chemical, etc. The non-conventional energy like nuclear energy related to atomic and sub-atomic particle forms has to be dealt with separately because in that case all matter would have to be considered as forms of energy as per the famous Einstein equation $E = mC^2$. The nuclear thermodynamics is altogether different branch and in order to distinguish the general thermodynamics the terms like chemical or metallurgical or mechanical thermodynamics have been introduced lately. Here the subject matter of discussion is chemical and/or metallurgical thermodynamics alone. The systems under discussion will be one of finite dimensions consisting of a large number of particle, *i.e.*, macro systems.

1.2 THERMODYNAMICS AND RATE OF CHANGE

The knowledge of thermodynamic science may indicate the possibility of the change but it does not indicate in any way the time required to bring about the change, *i.e.*, the rate of the change is not predictable by thermodynamics. The rates of changes are studied under a separate branch of science known as science of **'kinetics'** which helps to assess the time required for a certain change to occur *vis-a-vis* the physical configuration of the system. This can best be illustrated with the help of the following example :

The loss in potential energy or more appropriately the exchange of energy as the water flows under gravity is the same irrespective of whether the outlet is say 10 mm dia, 25 mm dia or inthe extreme a capillary. But the time required for a certain amount of water to flow down varies considerably in these cases.

In other words the rate of exchange of energy is different in these three cases. More appropriately the rate of thermodynamic change is different in these three cases and it depends on the physical nature of the system.

1.3 ENERGY

What is after all energy? It is the ability to do work, useful or otherwise. This is too mechanical an answer. If sulphuric acid and water are mixed, even gently without much mechanical work, heat is produced. Where was this thermal energy residing earlier? In the acid or in the water? Similarly if silicon is dissolved in iron huge amount of heat is produced. Where was it earlier that it is evolved on dissolution?

Having once established through the famous Einstein Equation $E = mC^2$ that even mass is a condensed form of energy, the whole universe is nothing but energy in its various forms, like mass, thermal, kinetic, chemical, magnetic, etc. On terrestrial level also it is true to the extent that everything is in terms of energy. Quite obviously the energy associated with different material forms is different and therefore when one form changes to another the energy either gets liberated or is absorbed in one or the other forms of energy. As a corollary energy may be used to bring about changes in material forms. **Natural changes in material forms are accompanied by evolution of energy, *e.g.*, burning of coal, whereas unnatural changes have to be brought-in by investing energy.** Natural changes are not many. Even many such have a history such that earlier it had been unnaturally brought into the form, like water stored in overhead tank, that now it flows down as natural change. Unnatural changes in material systems are necessary to produce so many materials for sustaining human civilization. Therefore nothing will change unnaturally from as-is-where unless some energy is invested to bring about that change.

Energy is therefore **capacity to bring about changes in the existing material forms as per requirements.** This definition is much broader than the mere ability to do mechanical work. Some material forms possess so much energy that given a chance they will undergo changes on their own, liberate excess energy and take a more stable, low energy level, material form in consequence.

Understandably such material forms are unstable because of higher energy associated with them. By and large naturally occurring material forms are not at high energy level and are processed on industrial level to produce a variety of material forms required for human civilization. These manufacturing activities therefore require energy. The industrial manufacturing activities are such that more than the theoretical energy is always required to carry out these industrial activities. This gives the energy efficiency of the manufacturing operation. This energy efficiency is one of the most important parameter that affects manufacturing economy and is under constant surveillance for improvement. The thermodynamic study is relevant from this point of view.

1.3.1 Energy Properties

The reader upto this stage is familiar with mass, pressure, temperature and volume as these are far less abstract ideas. One can feel or almost **'see'** these properties. Although energy exchanges during heating or cooling or phase changes can be assessed in terms of these simple properties, these properties by themselves are not adequate enough to assess energy exchanges during most of the changes. One of the simple reason for such an inadequacy is that these properties are not by themselves energy parameters as such. There must be some energy parameters or properties of materials such that during any change, and change they will, so that the energy exchanges would be assessable directly in terms of such energy parameters.

The thermodynamic science has evolved several such properties of materials which are energy parameters and by mere algebraic summation of their values before and after the change can reveal the possible energy exchanges in the changes. The following examples can be cited as illustrations.

1. A system takes some heat and in return does some work. Where is the balance energy? The thermodynamic science calculates this in terms of changes in **'internal energy'** which is a energy parameter.
2. A liquid is converted into vapour at the boiling point without changing the pressure and the temperature. The latent heat is provided for the change. The question is where and in what form the energy of vaporisation exists after vaporisation. The thermodynamic science evaluates this in terms of a property called - **'entropy'** of a system and which is a energy parameter.
3. A metal and oxygen combines to form the oxide. In what form the energy is lost from the materials? The thermodynamic science assesses it in terms of either **'enthalpy or free energy'** which are yet another energy parameters.
4. When electrolysis of a solution or an electrochemical reaction takes place it is thermodynamic science that can assess the energy exchange in terms of enthalpy and free energy parameters of the chemical reaction taking place.

1.3.2 Thermodynamic Laws

The observations on changes taking place in this universe have led to evolve some definite facts which were formulated in due course of time into specific statements known as the **Three Laws of Thermodynamics.** The **Zeroth Law** came up later. Although there is no direct proof to prove the validity of these thermodynamic laws but their applications to various changes taking place in this universe do yield correct results. These laws are applicable to finite changes in materials of finite dimensions consisting of large number of particles. These laws can not be applied to changes in materials of infinite dimensions such as on an universal scale or equally to micro-systems of just few atoms or molecules.

1.3.3 Forms of Energies

Since the thermodynamics deals with energies in all its forms it would be worth-while to know the possible forms of energies in material aggregates.

The concept of energy is foremost in modern science and it is the basic consideration in manufacturing processes from the point of view of process economy. Yet the concept of energy is not obvious. The concept of kinetic energy became apparent first, potential energy inalienably followed. In 1851 Lord Kelvin defined energy as, "**The energy of a material system is the sum expressed in mechanical units of work, of all the effects which are produced outside the system when the system is made to pass in any manner from the state in which it happens to be to a certain arbitrarily fixed initial, *i.e.*, 'standard state',**". It became obvious that there is nothing like absolute energy but only relative energy and that only energy changes can be measured. From the point of view of chemical and metallurgical engineers kinetic energy is seldom of interest. The total energy devoid of kinetic energy is called '**internal energy**'.

Energy has also been defined as capacity to do work. Work is considered to be done if a body moves under the influence of force applied on it. Therefore

$$w = f \times d$$

where f is force in dynes and d is the distance of movement. Work is measured in terms of erg in C.G.S. and Joule in S.I. units.

$$1 \text{ Joule} = 10^7 \text{ ergs}$$

In other words what is capable of doing work must possess energy. Energy manifests itself in variety of forms like thermal (heat), mechanical (in terms of movements), electrical and so on. Except the nuclear all other forms of energies and their interconversion and conversion to work specifically shall be dealt with in the text to follow.

Mechanical :

(i) kinetic energy = 1/2 × mass × velocity

(ii) potential energy = mass × height × acceleration component

(iii) configurational = integral force × distance over energy within certain limits

Thermal :

Heat exchanged = mass × specific heat × temperature difference

Electrical :

$$\text{Electrical energy} = \text{current} \times \text{time} \times \text{potential difference}$$

Chemical :

$$\text{Chemical energy} = \text{number of chemical bonds} \times \text{bond strength}$$

1.3.4 Sources of Energy

Although theoretically everything in this universe is energy, practically the resources of energy are limited. The naturally occurring materials which on mere initiation undergo changes into material forms, and thereby liberate energy during the natural change, are one of the chief sources of energy on the earth. The coals in all their forms, mineral oils and natural gas on burning give thermal energy which is either used directly or converted into electrical or mechanical energy. This is the chief source of energy and these materials are classed as chemical fuels. Fuel is considered to be synonymous for source of energy. Water stored, naturally or artificially by dams at high levels has potential energy which can be converted into electrical energy by hydel generation and hence is a source of energy.

These are conventional sources of energy. With the rise of modern civilization into a enormous manufacturing activity the energy requirement is ever increasing. The natural sources of energy are limited and hence attention is focused into **alternate sources of energy**.

In this category comes utilization of heat energy of sun-rays or kinetic energy of winds and tides, by converting them into suitable energy forms preferably electrical, mechanical or thermal. The bio-gas generation by disintegration of vegetables and bio-mass also comes in this category.

The nuclear energy available from fissionable materials like uranium was considered to be a very good potential source of energy on earth but the hazards associated with its waste disposal and the necessity of high safety standards has posed great hurdles in its full development as a source of energy. The initial euphoria in this field in the fifteens has now dampened considerably.

The advantages of chemical fuels as a source of energy is that it can be stored and carried conveniently and used as and when required. On the contrary electrical energy can be conveyed over hundreds of kilometers within no time but can not be stored readily. It must be used when available. Energy can be stored through chemical cells but the investment required is rather too large.

1.4 THERMODYNAMIC SYSTEM

It would be worth while to define certain terms before understanding the thermodynamic laws. Since the laws of thermodynamics are applicable to material aggregates of finite dimension, a well defined finite aggregate of materials has to be visualised or in reality constructed for the application of these laws and their derivatives for precise deductions. Such an aggregate of the material or a portion of this infinite universe, with real or imaginary dimensions, is referred to as a **'thermodynamic system'**, quite often for the sake of brevity only a **'system'**. In a dynamic bulk a portion of it can be imagined with finite imaginary boundaries as a system. The space or materials around this thermodynamic system under consideration is generally referred to as **'surroundings'**. The system may exchange matter or energy or both with the surroundings during the change. The thermodynamic systems can be classified based on its interaction with the surroundings or on its own material distribution or on its own composition as shown in Fig. 1.1

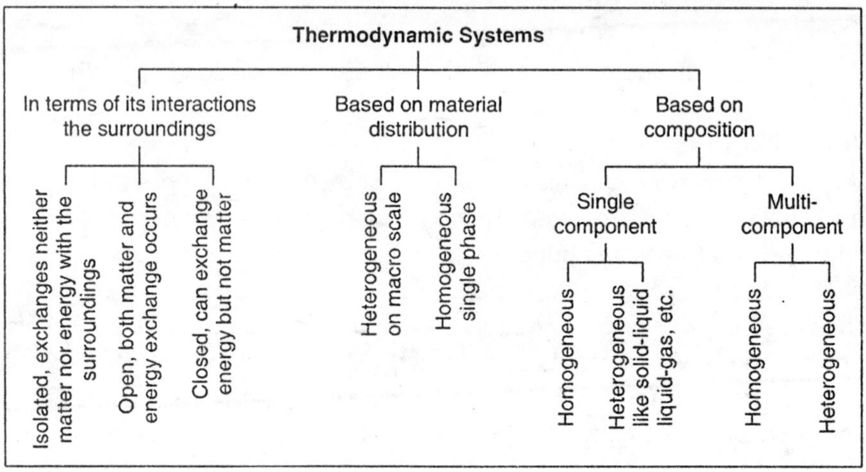

Fig. 1.1 Classification of thermodynamic systems.

A thermodynamic system must, of necessity, be stable with respect to its chemistry during its study. If inherently the system is undergoing continuously some chemical change it can not be considered as a thermodynamic system. For example a live animate body like tree or a human being are not systems because they are continuously undergoing some changes - the life sustaining processes. Similarly a closed circuit Daniel cell can not be a system because the anode and the cathode chambers are continuously undergoing chemical changes. On the other hand almost all of the inanimate aggregates of well defined boundaries can be called thermodynamic systems as their chemistry is fixed (systems in which extremely slow chemical processes are going on are exceptional). There can be some peculiar unstable chemicals which may not fall under the category of systems but these are exceptions. In short there must be **complete-internal-chemical-stability** for anything to be called as a thermodynamic system.

1.4.1 State of System

The system has to be described by some experimentally determinable parameters so that it can lead to its complete description and is understood by one and all as that particular system. For example for a single component homogeneous system therefore only two variables are to be fixed out of the three *viz.* pressure, volume and temperature for its complete description. Such variables are known as **'state'** variables. The minimum number of variables required to describe the state of the system are called **'independent'** state variables. It is also referred to as **'state properties'**.

The state of the system can also be described in terms of energies associated with it. The algebraic summation of the energy parameters of the system in two different states gives the energy exchanged during the change of the system from the initial to the final state.

1.5 EXTENSIVE AND INTENSIVE PROPERTIES

If a state variables, whether dependent or independent, is a function of the mass of the system it is known as **'extensive'** state property, *e.g.*, volume, energy properties, etc. It only means that these properties get added up in proportion to their mass or the size of the system and are expressed per unit mass like per gram mole. If however the function is independent of the mass or size of the system it is called **'intensive'** state property like density, temperature, pressure, conductivity, etc. These are not additive properties. For example when one gram of a substance at T degrees absolute is added to say a ton of similar material at the same temperature the resultant temperature is still T degrees only. Product of an intensive and an extensive properties is also an extensive state variable whereas the ratio of two extensive properties yields an intensive property. The values of extensive properties, expressed per unit quantity of the system or substance, become intensive variables like specific volume, specific heat, partial properties, etc.

1.6 EQUATION OF STATE

The state of the system can be described in the form of some mathematical equation involving some state variables.

This analytical form as applicable to the system under consideration is known as **'equation of state'**. For example the pressure, volume, temperature and mass of an ideal gas are interrelated by an expression as

$$PV = nRT \quad ...(1.1)$$

where n is the number of moles and R is the universal gas constant. This is the equation of state of ideal gas. Therefore if any two variables are fixed the third is automatically fixed for a given mass of ideal gas. The above relationship is an expression which correlates the pressure, temperature and volume of an ideal gas. In fact this is found to be true in case of solids as well as liquids though the exact form of this relationship may not be known. The same can be described in a generalised functional form as

$$F(P, V, T) = 0 \quad ...(1.2)$$

It only means that there exists a function for every substance in any form which correlates pressure, volume and temperature. It is not even remotely implied that it must be known. It may or it may not be known.

There can be several such relationships correlating the available state properties of a particular system and all such relationships shall be called as equations of state for that particular system. These are useful relationships to calculate the required state properties knowing some others.

1.7 THERMODYNAMIC PROCESS

A material system may undergo a change, under externally or internally imposed constraints, in terms of their state variables, from the existing one to some different variables. Such a change in the state of the system is known as a **'thermodynamic process'**. If a ten meter long rod is cut into two pieces the change is merely a physical one of breaking it into two pieces. But if in the process of breaking, the energy required to break the rod is taken into consideration then the process of breaking is considered to be a thermodynamic process. Similarly if iron combines with oxygen to form its oxide, it is a simple chemical process but if the energy exchanged in this process is taken into account then this process of oxidation is a thermodynamic process. The way the thermodynamic process takes place is known as its '**process path**'. There are different types of thermodynamic processes depending upon the conditions imposed on the system and are named accordingly and are shown in Fig. 1.2.

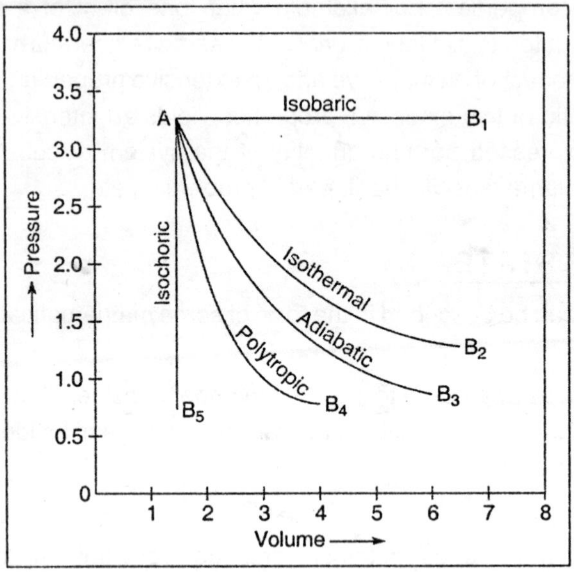

Fig. 1.2 Schematic representation of pressure vs. volume in various thermodynamic processes like isobaric, isothermal, isochoric, adiabatic and polytropic.

Isothermal - in which temperature remains constant and is mathematically described as $dT = 0$. It is imagined that the system is in contact with a thermal jacket maintained at the required temperature.

Isobaric - In which pressure does not change and hence $dP = 0$. It is imagined that the process is carried out in open atmosphere or in a vessel where the pressure is maintained constant.

Isochoric - In which volume does not change and hence $dV = 0$. It means that the process is carried out in a vessel of known volume.

Adiabatic - In which no exchange of heat occurs with the surroundings and hence $dq = 0$ and is described as PV as constant.

It is assumed that the process is insulated from the surroundings perfectly. The P or V may change as per the given equation.

Polytropic - Those processes which obey equation PV is constant where PV is any positive number between one and γ but not equal to either of them. In this all variables can change but since any two can determine the third therefore, only two vary and the popular form of this equation therefore takes the form involving P and V only.

1.8 PATH AND STATE FUNCTIONS

It is easier to visualise that the change in pressure, volume or temperature during a thermodynamic process is entirely related to the original state of the system and the final state to which it has changed to. It does not depend on which way or by what process path the change has actually occurred. For example if a gas is expanded from say a volume of 10 meter cube to 15 meter cube by an adiabatic or isothermal change, the change in volume is obtained by algebraic summation of the two end states and it is 5 meter cube in this case. The same is true of temperature and pressure. he property whose change depends on only the initial and the final states of the system and not on the path adopted to bring about the change, is called a **'state function'**. Mathematically therefore if the property is a state function then in a cyclic process, when the system undergoes a change but returns back to its original state,the net change in the state property or function is zero. If x is the state function then

$$\oint dx = 0 \qquad ...(1.3)$$

and if say γ is not a state function then $\oint d\gamma$ need not necessarily be equal to zero, where $\oint$ indicate the integral over the entire cyclic change. As against this there are properties such that the change in them in a process depends on the path followed for this change. It is interesting to note that energy is compounded of two terms namely, heat and work effects produced outside the system, neither of which are uniquely related to the state of the system.

Let us consider a gas being compressed or expanded from state A to state B as shown in Fig. 1.3. It can be changed from state A to state B, *i.e.*, from configuration P_1V_1 to P_2V_2 by any of the courses 1, 2 or 3. The work done in such a case shall be the area under the relevant curve from A to B, and it shall be integral PV. In course 1 the process is carried out isothermally. In course 2 it is first by cooling followed by heating and in course 3 it is first by heating followed by cooling. The work done in all these cases is different though the initial and final states are exactly the same, because of different paths followed in these cases. Therefore the work done by making the system change from state A to B is not a function of states A and B but it definitely depends upon the path adopted for the change. Similarly the ability to absorb or desorb heat, *i.e.*, heat exchanged in the process depends on the path followed. Hence work done **w** or heat exchanged **q** are not state functions but path functions.

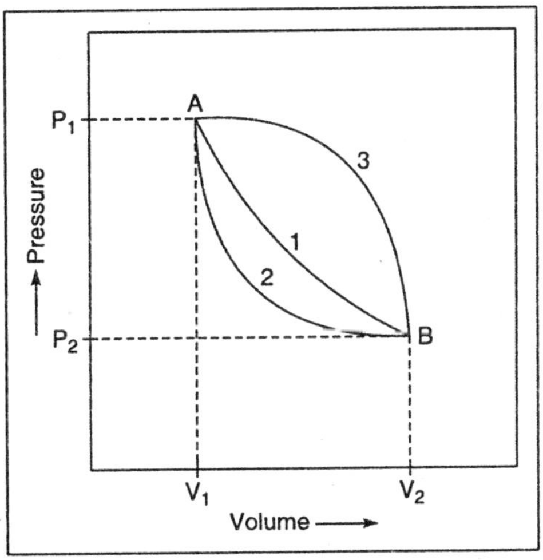

Fig. 1.3 Schematic P-V diagram showing the expansion and contraction of a gas between state A depicted by P_1V_1 and state B depicted by P_2V_2.

The usefulness of state functions is also obvious from the above example because mere algebraic summation of the values of the state functions gives the change in the state but in this instant case of assessing the work and heat exchange effects such a summation does not lead to the required value as these are not state functions. Whether any function is a state function or not can be mathematically assessed as follows. For an ideal gas T = PV/R wherein P and V are taken as independent variables and T as dependent variable. The above can be expressed in its total differential form as

$$dT = \left(\frac{\partial T}{\partial V}\right) dV + \left(\frac{\partial T}{\partial P}\right) dP \qquad ...(1.4)$$

generalised form if a system has two independent variables say **x** and **y** and any third function **z** can be expressed in its total differential form as

$$dz = M\,dx + N\,dy \qquad ...(1.5)$$

where M and N may be functions **x** and **y** hence function **z** shall be a state variable if and only if

$$\frac{\partial M}{\partial y} = \frac{\partial N}{\partial x} \qquad ...(1.6)$$

Whether a function is a state function or not can thus be established by relating it with two other known state functions through the above relationship and if it comes out to be true it shall be a state function.

1.9 THERMODYNAMIC EQUILIBRIUM

This is very important concept for understanding the usefulness of thermodynamic science and its applications. This can be readily understood with the help of common examples. Let us imagine a inflated tube of an automobile vehicle. If it is punctured in an open atmosphere the pressurised air from the tube leaks out through the puncture like a jet till there is no extra pressure left inside the tube. What will happen if the same inflated tube is punctured in a chamber which has also been inflated to the same pressure ? The tube will not get deflated. Since pressure inside the tube and outside are equal, *i.e.,* since mechanical potentials on either side of the tube wall are the same, there exists a **'pressure or mechanical equilibrium'** in the system - the tube and its surrounding chamber.

Similarly, if two objects at different temperatures are physically joined heat flows from the one at higher temperature to the one at lower temperature. If the two were at the same temperature no heat flow shall occur. It is then said to be in **'thermal equilibrium'**.

If two phases are in contact such that a chemical reaction takes place between them, but the two phases can be present in such a way that the rates of forward and the backward reactions are exactly matched then the system can be called as in **'chemical equilibrium'**. The chemical potentials in both of these phases must be the same in that case.

Complete **'thermodynamic equilibrium'** is thus that situation where the system is in equilibrium with respect to all such potentials like mechanical, thermal, chemical, etc. The term equilibrium is applied to a system and not to a process.

1.9.1 Partial Equilibrium

A system can be in a state of partial equilibrium. It only means that the two phases involved in a system are in equilibrium with respect to any one of the above mentioned parameters, thermal, chemical, etc. or a combination of these but not all of them at a time. If a system contains a semi-permeable membrane separating say a mixture of nitrogen and hydrogen gases on one side and only hydrogen on the other, it is likely that the partial pressures of hydrogen on either side may be equal because of the membrane being permeable to hydrogen, but the total pressures may be different because of the nitrogen being impermeable through the membrane. It may take ages before the system has total pressures equal on both the sides of the membrane. Strictly speaking the system is not truely a thermodynamic system and hence not amenable for thermodynamic treatment using state variables because of the continuously changing composition on either side of the membrane. Yet within a reasonable time scale the system can be considered to be a system and treated accordingly. The equilibrium in such a case is partial equilibrium and can be so treated with negligible error involved in the calculations.

1.10 REVERSIBLE AND IRREVERSIBLE PROCESSES

Thermodynamic processes can also be subdivided as reversible and irreversible processes. These terms are applied to a process involving a system but not to a system as such.

A process that can be reversed in its direction by an infinitesimal change in the driving force, *i.e.,* change in one or more of the state variables is said to be a **'reversible process'**. A classical example of this is the gas cylinder and a piston. If the pressure of the gas is say P atmospheres and if (P + dP) is exerted from outside on the piston, the gas inside the cylinder shall be compressed. However, if the external pressure is (P – dP) then the gas shall expand. A classical chemical reaction like

$$A + B = C + D$$

if can be made to go in the backward or forward direction by changing one or more of the variables, then it is a reversible reaction. A Daniel cell has an open circuit emf of 0.76 volts and if the circuit is closed then current flows in the direction of zinc to copper in the solution and copper to zinc in the electronic conductor. If an opposite external emf is applied the current slowly decreases and is zero when exactly the same value of emf is superimposed on the cell in opposite direction. Any further increase now would make the current flow from copper to zinc in solution, *i.e.,* reverse. Such a process can be made to go in the desired direction by changing the superimposed emf. This is a typical reversible process.

On the contrary a match stick when it burns, the process can no be reversed by changing one or the other parameters. Once burnt, the stick can not be reproduced by reversing the process. This is typically a **'irreversible process'**. Other examples are mixing of two gases, mixing of two liquids to form a solution or flow of electric current through a resistor and so on. All natural processes are most often irreversible changes.

If the case of a gas-cylinder and a piston is taken as an experimental tool, it can be shown that the work done in expanding the gas very slowly through a volume change V_1 to V_2 is given by

$$w_{1 \to 2} = \int_{V_1}^{V_2} P\,dV \qquad ...(1.7)$$

and that to get the system back to its original position is

$$w_{2 \to 1} = \int_{V_2}^{V_1} P\,dV = -\int_{V_1}^{V_2} P\,dV \qquad ...(1.8)$$

If the process is carried out not so slowly then it can be proved that the work done in similar expansion and contraction is less than that done when the process is carried out slowly. Hence

work (reversible) > work (irreversible)

This is so because in reality the pressure difference driving the system to do work is maximum in reversible state.

Although this is a simplified picture under isothermal conditions applicable to ideal gas it holds true for any working substance gas, solid or liquid under any conditions.

1.11 CONCLUDING REMARKS

The enormous industrial manufacturing activities required to sustain the present-day civilization requires equally enormous energy. The energy resources are limited. It is therefore necessary to understand the scientific principles underlying energy requirements in these manufacturing activities so that the minimum energy input may be achieved. The thermodynamic laws and their applications contribute towards this understanding. The basic terms such as energy, system, process, state and path properties, reversible and irreversible processes, extensive and intensive properties and so on are required to understand these thermodynamic laws and these are therefore covered in the present chapter and the details of the thermodynamic laws are covered in the following chapters.

□□□

Chapter 2

First Law of Thermodynamics

Internal Energy, Enthalpy and their Derivatives

2.1 HISTORICAL

As seen earlier, from the definition, the measurable energy of any system is dependent on a arbitrarily chosen standard state. These is nothing like measurable absolute energy of a system, it is all relative. Thus only energy differences can be measured. A system may have huge amount of energy but in a given surrounding it will be possible to give out only a certain amount of energy and that will be the energy exchange possible and it will be measured only in terms of the energy so released. In the present discussion and as is of no consequence, the kinetic energy and the mass as a form of energy in nuclear terms, are omitted, being constant in any process, unless otherwise stated.

It was Rumford who first appreciated the conversion of work into heat in qualitative manner in 1789. This was more properly explained by H. Davy in 1799 by showing that ice melted when two ice blocks were rubbed against each other under vacuum. It was however Joule who established through his famous experiment, in an adiabatic enclosure, the quantitative equivalence of work and heat as

$$\frac{\text{work}}{\text{heat}} = \text{Joule} \quad \text{or} \quad J = \frac{w}{q} \qquad \text{...(2.1)}$$

If both work and heat are expressed in the same units then for a cyclic process

$$w = q \qquad \text{...(2.2)}$$

It means that energy in the form of work gets transformed into another form of energy as heat fully. This formed the basis of the law of **'conservation of energy'** which is the progenitor of the **first law of thermodynamics**. It also means that it is just not possible to produce energy of any kind without dispensing an equivalent amount of energy of some other form. It is therefore impossible to construct a **'perpetual machine'** which will do work perpetually without putting energy into it in any other form.

2.2 FORMULATION OF FIRST LAW OF THERMODYNAMICS

In a gravitational field work done by a body as a result of its movements depends on its initial and final positions and not on the path taken by the body between these two positions. Similarly the work done by a charged particle under the influence of electric field depends on the potential difference between the two points, between which the particle moves, rather than on the path taken. Thus the work done, in the above, by a body in its movements from state **A** to state **B** depends on the functions of states **A** and **B** and not on the path the body takes for its movements.

In other words the work done by a body in an adiabatic enclosure when moved from its initial to final state involves the same amount of work and which is independent of the path followed for the change. The change in the internal state of the system, as a consequence of which the work is done, has to be defined and is called the **'internal energy'** of the system denoted in the literature by either letters **'U'** or **'E'**. Here it will be referred to as **'U'**.

In an adiabatic system the work done by the system on the surroundings shall be equal to the change in the internal energy of the system as

$$\Delta U = w \qquad \text{...(2.3)}$$

Similarly if a system while doing work on the surroundings simultaneously accepts heat from the surroundings then the work done shall be equal to the change in the internal energy of the system as well as the surroundings. The change in the internal energy of the surrounding shall be equal to the heat exchanged. Hence

$$q = \Delta U' \text{ (surr)} \qquad \text{...(2.4)}$$

If the system and the reservoir supplying heat are considered as a single system and if this is isolated adiabatically, then the work done by this combined system equals the change in the internal energy of the combined system, or

$$w = (\Delta U' - \Delta U) \qquad \text{...(2.5)}$$

putting Eq. (2.4) into this gives

$$w = -\Delta U + q \qquad \text{...(2.6)}$$

or on an infinitesimal form

$$dU = \partial q - \partial w$$

This is nothing but mathematical statement of the **'First Law of Thermodynamics'**. In a generalised form it can be said that if ∂q amount of heat is supplied to a system and if it does work ∂w on to the surroundings the change in the internal energy of the system is given by Eq. (2.6).

2.3 INTERNAL ENERGY AS A STATE FUNCTION

Let us go back to Fig. 1.2. The system happens to be in a state **A** and undergoes a change to state **B**. Firstly it follows the path 1 and second time it follows the path 3. In both the cases the system returns to its original state via path 2. In these two independent cyclic processes the net change in energy is zero as the system returns back to its original state. This has to be. The heat exchanged and work done in each cycle will have to be equal. Even from Eq. (2.6) if energy change is zero q has to be equal to w.

Therefore heat exchanged and work done in each cycle can be equated as in Eq. (2.2) as

$$w_2 + w_1 = q_2 + q_1 \qquad \text{...(2.7)}$$

and

$$w_3 + w_2 = q_3 + q_2 \qquad \text{...(2.8)}$$

or $$q_2 - w_2 = w_1 - q_1 \quad ...(2.9)$$

and $$q_2 - w_2 = w_3 - q_3 \quad ...(2.10)$$

and hence for that matter for any path followed for a cyclic change with state A and state B as initial and final states

$$q_1 - w_1 = \text{constant}$$

or when expressed in infinitesimal form as

$$\partial q - \partial w = \text{constant} \quad ...(2.11)$$

Therefore for any change the actual values of w and q shall depend upon the path followed but the difference shall always be the same as long as states A and B as initial and final states remain the same. In other words the difference (q – w) shall not depend on the path, although individually both of them depend on the path followed, but on the states A and B only and, therefore (q – w) shall be a **'state function'**. For U to be a state function the values (q – w) has to be zero for a cyclic process and which is the case in Eqs. (2.7) and (2.8). **Internal energy is therefore a state property.**

It reveals that the system is capable of absorbing heat from the surrounding and doing work on to the surrounding such that Eq. (2.6) holds true. It is quite possible that the system may, as a whole, result in positive or negative change in U depending upon the actual values of heat q received and the amount of work w done.

Hence sum of all forms of energies exchanged by a system with the surroundings is equal to the change in its internal energy. It can also be stated as that **'energy can not be created nor can it be destroyed'**. This is one of the statements of the first law of thermodynamics. It can also be stated as **'the total energy of a system and its surroundings shall be constant and that it may change its forms or may exchange between themselves'**.

The system is therefore capable of accepting energy in one form and deliver it in the same or different form and in the same or different amount. This is possible because the system is capable of acting as a **'reservoir'** of energy. The stored energy is nothing but the internal energy of the system referred above and it increases or decreases depending upon its interaction with the surrounding or changes within itself.

The internal energy consists of all types of energy forms contained by the system including mass if nuclear reactions are under consideration. As per Kelvin's definition the presence of any particular form of energy will be seen only as a manifestation that the system produces, during its interaction with the surroundings. For example if nuclear reactions are not being considered mass will not manifest itself as energy. Similarly if electrical effects are not considered electrical form of internal energy will not manifest. In chemical and metallurgical processes the thermal and mechanical energies alone manifest and therefore internal energy in those forms alone shall be manifest and would be considered hereafter, unless otherwise stated specifically since chemical and metallurgical thermodynamics is being considered here.

So far the reader is conversant with the terms, volume, pressure and temperature as state properties and how internal energy U has been added to this list of state properties to describe the energy content of the system. The question is whether there is any correlation between the directly measurable state properties like P, V and T and the newly coined property U?

If the system expands by a volume change dV against pressure P then the mechanical work done by the system on to the surroundings is $\int P\,dV$ and the Eq. (2.6) shall be modified as :

$$dU = dq - P\,dV - w' \quad ...(2.12)$$

where PdV is the mechanical work and w′ is all other forms of work like electrical, magnetic, surface, etc. In normal chemical and metallurgical systems w′ is almost always zero because the system does only mechanical work and therefore henceforth w′ is assumed to be zero, unless otherwise stated specifically. Therefore

$$dU = dq - PdV \quad ...(2.13)$$

It is desirable to have equation involving U the internal energy in the form where all quantities are exact differentials, *i.e.*, in terms of state properties. But this is the best first law of thermodynamics can lead to. This can be cast in the desired form only with the help of second law of thermodynamics.

2.4 INTERNAL ENERGY IN TERMS OF PARTIAL DERIVATIVES

It has already been stated that for homogeneous system consisting of a given amount of substance of fixed composition the equation of state of that system can be described as

$$F(P, V, T) = 0 \quad ...(2.14)$$

and since out of P, V and T only two are independent variables, the third gets automatically fixed and it can be described as

$$P = F(V, T); \quad V = F(P, T) \quad \text{and} \quad T = F(V, P) \quad ...(2.15)$$

The internal energy U being a state function will have to be a function of P, V and T and that is to say

$$U = F(P, V, T) \quad ...(2.16)$$

This also therefore can be stated as (like in Eq. 2.15)

$$U = F(P, V); \quad U = F(T, P) \quad \text{and} \quad U = F(V, T) \quad ...(2.17)$$

and from the theorem of partial derivatives dU can be expressed as

$$dU = \left(\frac{\partial U}{\partial P}\right)_V dP + \left(\frac{\partial U}{\partial V}\right)_P dV \quad ...(2.18)$$

$$dU = \left(\frac{\partial U}{\partial P}\right)_T dP + \left(\frac{\partial U}{\partial T}\right)_P dT \quad ...(2.19)$$

$$dU = \left(\frac{\partial U}{\partial V}\right)_T dV + \left(\frac{\partial U}{\partial T}\right)_V dT \quad ...(2.20)$$

and from these under the conditions of isothermal, isobaric or isochoric processes the dependency of U on the state variables can be established. These equations in which internal energy is expressed as a function of independent state variables are called **'caloric equations'** of state as these allow evaluation of thermal capacities or heat capacities. For a single component system the caloric equation can be expressed also as

$$F(U, T, V) = 0 \quad ...(2.21)$$

and for multi-component systems it will be

$$F(U, T, V, n_1, n_2, n_3, \ldots, n_i) = 0 \qquad \ldots(2.21a)$$

2.5 HEAT CAPACITY

In any process, but more so in chemical and metallurgical processes, materials get heated or cooled and therefore it is necessary to know the amount of heat required to heat or as a corollary amount of heat liberated, in cooling a material, over a certain temperature range. Different materials require different amounts of heat to get heated through the same temperature rise. this is because the materials have different heat capacities. This is so because of the variation in the crystal structure of materials and their related parameters.

The heat capacity of a substance is the amount of heat required to raise its temperature by one degree. In a limiting case the same can be expressed as

$$C = \frac{\partial q}{\partial T} \qquad \ldots(2.22)$$

The term specific heat refers to the heat capacity of unit mass of the material under consideration, *i.e.*, it is that heat required to raise the temperature of a mass of a gram of substance through one degree. Since q depends on path the value of heat capacity may also depend on path, if expressed in the form mentioned as above. In an adiabatic process when q is zero the factor q/dT becomes zero, which is absurd. It is necessary therefore to express heat capacity in terms of state functions.

As internal energy is related to q and as it is also related to state functions, the heat capacity, through q could be related to state functions. It should also be noted that the heat capacity depends on conditions of constancy of volume or pressure, particularly in the case of gases, perhaps less so but definitely as well in the case of liquids and remotely so for solids as well. Hence two different heat capacities have been estimated, one as C_V as heat capacity at constant volume and the other C_P as that at constant pressure.

2.5.1 Specific Heat at Constant Volume

From first law of thermodynamics as in Eq. (2.13)

$$dU = \partial q - P\,dV$$

or

$$\partial q = dU + P\,dV$$

substituting in this the value of dU as given in Eq. (2.20)

$$\partial q = \left(\frac{\partial U}{\partial V}\right)_T dV + \left(\frac{\partial U}{\partial T}\right)_V dT + P\,dV$$

$$= \left(\frac{\partial U}{\partial T}\right)_V dT + \left[\left(\frac{\partial U}{\partial V}\right)_T + P\right] dV \qquad \ldots(2.23)$$

when dV = 0 under constant volume conditions

$$\partial q = \left(\frac{\partial U}{\partial T}\right)_V dT$$

$$\frac{\partial q}{\partial T} = \left(\frac{\partial U}{\partial T}\right)_V \quad \text{or} \quad \frac{\partial q}{\partial T} = \left(\frac{\partial U}{\partial T}\right)_V$$

or
$$C_V = \frac{\partial U}{\partial T} \quad \text{...(2.24)}$$

Thus **heat capacity under constant volume is nothing but rate of change of internal energy with temperature.** The same thing can be arrived at in a different way as follows :

Under the conditions of $dV = 0$ no mechanical work is done by the system and hence $w = P\,dV = 0$ and hence from the first law equation

$$(\partial q)_V = (dU)_V$$

or
$$\left(\frac{\partial q}{\partial T}\right)_V = \left(\frac{\partial U}{\partial T}\right)_V$$

or
$$C_V = \left(\frac{\partial U}{\partial T}\right) \quad \text{as shown in Eq. (2.24)}$$

Although q is a path function **heat exchanged under constant volume conditions is a state function** and equals change in internal energy.

2.5.2 Specific Heat at Constant Pressure

If pressure is maintained constant during change of a system from initial state (i) to final state (f) the work done will be

$$w = \int_i^f P\,dV = P\int_i^f dV = P(V_f - V_i)$$

The first law equation as in Eq. (2.6) can be rewritten as

$$U_f - U_i = dq - P(V_f - V_i)$$

or on rearranging these terms

$$(U_f + PV_f) - (U_i + PV_i) = q$$

The expression (U + PV) is referred to as **'heat content'** or **'enthalpy'** of the system and is often designated by the letter **H** as

$$H = U + V \quad \text{...(2.25)}$$

or thus
$$Hf - Hi = \Delta H = dq \quad \text{...(2.26)}$$

dividing by dT will give

$$\left(\frac{\partial q}{\partial T}\right)_P = \left(\frac{\Delta H}{dT}\right)_P \quad \text{or} \quad C_P = \frac{dH}{dT} \quad \text{...(2.27)}$$

The **heat capacity at constant pressure is nothing but rate of change of enthalpy with temperature.**

The same can be derived using Eq. (2.19) and substituting it in Eq. (2.13) as

$$dq = dU + P\,dV$$
$$= P\,dV + \left(\frac{\partial U}{\partial P}\right)_T dP + \left(\frac{\partial U}{\partial T}\right)_P dT$$

as $dP = 0$ $$dq = P\,dV + \left(\frac{\partial U}{\partial T}\right)_P dT$$

dividing by dT gives

$$\left(\frac{\partial q}{\partial T}\right) = C_P = P\left(\frac{\partial V}{\partial T}\right)_P + \left(\frac{\partial U}{\partial T}\right)_P$$

being exact differentials $\partial U = dU$ and $\partial T = dT$

and hence $$C_P = \left[\frac{d(U + PV)}{dT}\right]_P \quad ...(2.28)$$

substituting H for (U + PV) in the above

$$C_P = \left(\frac{dH}{dT}\right)_P \quad \text{as in Eq. (2.27)}$$

Alternatively the same can be derived as : From the mathematical

definition of enthalpy $H = U + PV$

differentiating gives $dH = dU + P\,dV + V\,dP$

and since from first law $dq = dU + P\,dV$

hence $dH = dq + V\,dP$

and at constant pressure since $dP = 0$

$dH = dq$

or $\Delta H = dq$ as arrived in Eq. (2.26)

and by dividing this by dT Eq. (2.27) is obtained.

Since enthalpy is a state function the heat exchanged by a system at constant pressure becomes a path independent property. In other words it is a state property. Also by definition itself H depends on all those parameters which are by themselves state properties and it itself, in the derived form, is a state property. Since enthalpy is a state property it can be expressed as a function of any two variables out of the three P, V and T and thus expressing it as a function of P and T as

$$H(P, T) = 0 \quad \text{or} \quad H = F(P, T) \quad ...(2.29)$$

2.6 INTERNAL ENERGY VS ENTHALPY

One can readily understand the term internal energy as all that energy stored in the system. But it is rather difficult to understand the term enthalpy because it is defined by mathematical equation. Therefore **what does enthalpy mean?** This can be best illustrated by the following example :

Let us imagine the process of calcination of calcium carbonate given as

$$CaCO_3 = CaO + CO_2 \quad ...(2.30)$$

Heat is supplied to break the chemical bonds between CaO and CO_2. Assume for the time being that both CaO and CO_2 are solids. Then the invested bond energy would be internal energy of the process of calcination. In this case the original system of $CaCO_3$ with say conditions as P, V and T remain practically the same after calcination. It would amount to merely a solid/solid phase change.

However if CO_2 is allowed to form as gas in this process then on breaking of one gram mole of $CaCO_3$ nearly 22.4 litres of CO_2 at STP will be formed as shown in Fig. 2.1. There is a tremendous

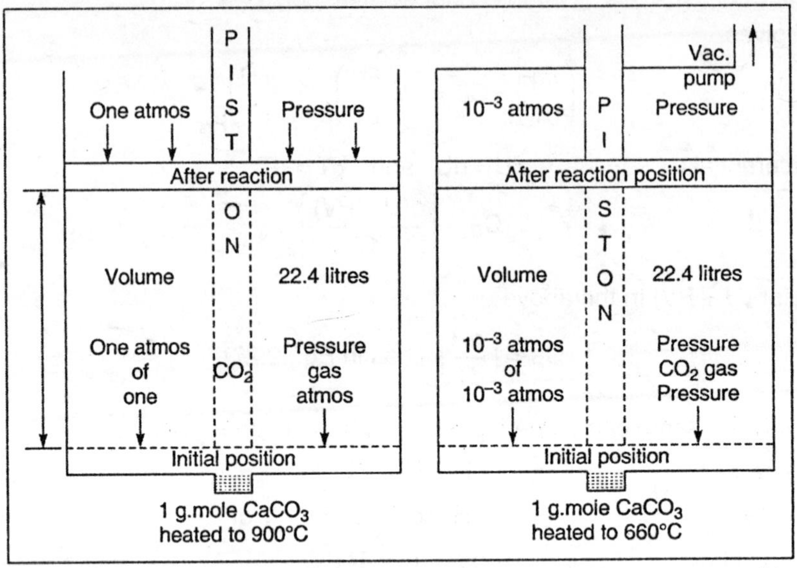

Fig. 2.1 Work done by the process of calcination of $CaCO_3$ under ambient pressures of 1 atmos and 10^{-3} atmos at two different calcination temperatures. Note the volume change after the reaction.

increase in the volume of the system on calcination. The system must do work of expansion to attain the resultant expected system. The process is supposed to be carried out at one atmospheric pressure in normal atmosphere. If the system has to attain a volume expansion of the order of 22.4 litres at STP then it should first attain a positive pressure against the ambient atmospheric pressure and then expand. In this process of expansion, the system will do work equivalent of $\int P\,dV$ as mechanical work on to the surrounding. Hence in addition to the requirement of energy for breaking the chemical bond of CaO - CO_2, additional energy equivalent of $\int P\,dV$ will have to be supplied to the system for this process making a total of U + P dV energy and this is the enthalpy change of the system on calcination, quite different from the internal energy U.

For chemical processes where there is no perceptible change in volume as in

$$2CaO\,.\,SiO_2 = 2CaO + SiO_2$$

the $\int P\,dV$ term is practically absent and hence U and H are almost the same. The difference in U and H arises because of gaseous component in the system. If there is no gaseous component and only condensed phases are present as in the equation above, there will be no difference between U and H. When condensed phase changes to another condensed phase, the change in volume is negligible or can be neglected or $\int P\,dV$ term shall be treated as zero. The perceptible value for $\int P\,dV$ arises only because of gaseous component involved in the reaction and, hence the difference in values of U and H.

A system thus has internal energy and as distinct from this may have enthalpy also. In chemical and metallurgical world therefore the enthalpy is often more relevant and distinctly different from internal energy. Even if the term $\int P\,dV$ is absent it is worthwhile to refer to H and forget about the reality as it takes into account what it should.

The usage of enthalpy changes as a measuring parameter is obviously more appropriate to take into account the work term of the process rather than mere internal energy. The usage of enthalpy is thus very popular rather than internal energy in chemical and metallurgical world. The term **'heat of reaction'** is restricted to the exchange of heat during a chemical reaction and which does not take into account the mechanical work term associated with it as explained above.

2.7 INTERRELATIONSHIP OF C_P AND C_V

From the first law of thermodynamics at constant pressure

$$dq = dU + P\,dV$$

and substituting the value of dU as given in Eq. (2.20) gives

$$dq = \left(\frac{\partial U}{\partial V}\right)_T dV + \left(\frac{\partial U}{\partial T}\right)_V dT + P\,dV$$

and dividing by dT gives

$$\frac{\partial q}{dT} = \left(\frac{\partial U}{\partial V}\right)_T \cdot \left(\frac{\partial U}{\partial T}\right) + \left(\frac{\partial U}{\partial T}\right)_V + P\frac{dV}{dT}$$

assuming constant pressure conditions

$$\left(\frac{\partial q}{\partial T}\right)_P = C_P \quad \text{and} \quad \left(\frac{\partial U}{\partial T}\right)_V = C_V$$

it will reduce down to

$$C_P - C_V = \left(\frac{\partial U}{\partial V}\right)_T \frac{dV}{dT} + P\left(\frac{dV}{dT}\right)_P$$

$$= \left(\frac{dV}{dT}\right)_P \cdot \left[\left(\frac{\partial U}{\partial V}\right)_T + P\right]$$

For an ideal gas the rate of change of internal energy with volume at constant temperature is nil. That is for an ideal gas because of its non-interactive nature.

$$\left(\frac{\partial U}{\partial V}\right)_T = 0$$

Hence $$C_P - C_V = \left(\frac{dV}{dT}\right)_P . P \qquad \text{...(2.31)}$$

For ideal gas $$PV = RT$$

or $$V/T = R/P$$

(when V changes with T, obviously P gets fixed up)

and therefore $dV/dT = R/P$ and substituting in Eq. (2.31) gives

$$C_P - C_V = R \qquad \text{...(2.32)}$$

2.8 CONSEQUENCES OF FIRST LAW OF THERMODYNAMICS

In the light of first law of thermodynamics various thermodynamic processes can be studied to derive additional relationships. The system studied is ideal gas and for which

$$PV = RT \quad \text{and} \quad (\partial U/\partial V) = 0$$

2.8.1 Isothermal Process

The process is carried out when dT = 0. The heat supplied to the system in this case must fully come out in the form of work done on to the surrounding. Hence

$$\partial q = \partial w \quad \text{or} \quad dU = 0 \qquad \text{...(2.33)}$$

It is also true that in this case

$$\partial q = \partial w = \int_{V_i}^{V_f} P\, dV = \int_{V_i}^{V_f} \frac{RT}{V} dV = RT \int_{V_i}^{V_f} \frac{dV}{V} = RT \int_{V_i}^{V_f} d \ln V \qquad \text{...(2.33a)}$$

Similarly,

$$\Delta H = \Delta U + \int PV$$

$$= \Delta U + (P_f V_f - P_i V_i)$$

$$= \Delta U + R(T_f - T_i) \qquad \text{...(2.34)}$$

Since there is no temperature change

$$T_f - T_i = 0 \quad \text{and combining with Eq. (2.33)}$$

$$\Delta H = 0 \qquad \text{...(2.34a)}$$

Hence in an isothermal process both internal energy and enthalpy remain unchanged and work is done by the system equivalent to the heat input. Therefore now for an isothermal process not only dT = 0 but dU = 0 and dH = 0 and reversible work done by the system is

$$q = w = RT \ln (V_f/V_i) \qquad \text{...(2.35)}$$

2.8.2 Isochoric Process

Under constant volume conditions the system does no mechanical work on the surrounding and hence $\int P\, dV$ equals zero. The first law Eq. (2.13) gets reduced to

$$dU = dq \qquad \text{...(2.36)}$$

and also

$$dH = dU + P\, dV = -dq \qquad \text{...(2.37)}$$

In other words the heat absorbed or rejected by the system during the process equals the change in the internal energy or enthalpy of the system. Therefore, for a constant volume process dV = 0 and also dU = dH = dq.

2.8.3 Adiabatic Process

There is no exchange of heat between the system and the surrounding, *i.e.*, dq = 0. Hence first law takes the form

$$dU = -dw = -P\, dV \qquad \text{...(2.38)}$$

For an ideal gas

$$dU = C_V\, dT = -P\, dV \qquad \text{...(2.39)}$$

Also since by definition

$$H = U + PV$$

therefore, $dH = dU + P\,dV + V\,dP$

and combining with Eq. (2.38) $dH = V\,dP$

That the system does work at the expense of its own energy and therefore temperature of the system shall fall. Using these relationships for a adiabatic process the interrelation of P and V or P and T can be found as

$$PV^{\gamma} = \text{constant} \quad \text{and} \quad TP^{\left(\frac{\gamma-1}{\gamma}\right)} = \text{constant}$$

where γ is constant. The reader will recall now how the relationship for adiabatic process was given earlier in Chapter 1, Section (1.7). Expressing dU in terms of partial derivatives as in Eq. (2.20)

$$dU = \left(\frac{\partial U}{\partial V}\right)_T dV + \left(\frac{\partial U}{\partial T}\right)_V dT = -P\,dV \quad \text{and } dq = 0$$

and since $$\left(\frac{\partial U}{\partial T}\right)_V = C_V \quad \text{and} \quad \left(\frac{\partial U}{\partial V}\right)_T = 0$$

for an ideal gas and also assuming C_V to be constant over a certain temperature range it leads to

$$C_V\,dT = -P\,dV$$

Adiabatic work done can be obtained by integration of $C_V\,dT$ over T_i initial and T_f as final temperature as

$$\text{Adiabatic work done} = -C_V\,(T_f - T_i) \quad \text{...(2.40)}$$

and since $dU = -w = -P\,dV$

adiabatic energy change $= C_V\,(T_f - T_i)$ ...(2.41)

2.8.4 Polytropic Processes

Since the polytropic processes obey PV^n = constant, the final expression of dU and dH accompanying the polytropic processes is quite complex because the mechanical work done is $\int PV$ and not $\int P\,dV$. Since both P and V change during the process, the work done by the system will have to be calculated taking changes in both into account and this is quite complex and therefore, not introduced here at the beginner's level.

2.8.5 Throttling and Free Expansion

This is a irreversible process carried out adiabatically as there is no chance for heat exchange to occur in such a instantaneous process of drop in pressure and free expansion in volume.

Since $dq = 0$, $dU = -dw$ and hence the system performs work at the expense of its own energy and thereby the temperature of the system shall fall. Therefore, if there is a jet coming from a nozzle in an open atmosphere the fluid temperature in the jet is lower than the temperature of the fluid fed to the jet.

2.8.6 Deviation of Real Gas from Ideality

The ideal gas equation correlating P, V and T is applicable for ideal gases. In reality the actual volume of gas is not the apparent one but, that minus the actual volume occupied by the gas particles (molecules). Hence **Clausius** modified it to

$$P(V - b) = RT \quad ...(2.42)$$

where '**b**' represents the volume occupied by the molecular particles of the gas as such. The real behaviour is further deviated from ideality because of molecular interaction and hence **van der Waal** further modified it to :

$$\left(P + \frac{a}{V^2}\right) \times (V - b) = RT \quad ...(2.42a)$$

This is known as **van der Waal's Equation** and it comes more closer to real observed behaviour of gases. This is a cubic equation for volume with three alternative values. When all the three values are real, gas can be liquified. The lowest temperature at which this is possible is called the **'critical temperature'** and above which gas will not be liquified. The relevant critical volume and critical pressure are the ones when liquefaction will be possible.

Although the modified form of gas equation to correlate the observed behaviour of real gases, is not a direct consequence of the first law of thermodynamics, it is explained here to understand limitations of ideal gas and its use in deriving certain expressions as in Eq. (2.32) in the form of derivations from first law of thermodynamics.

2.9 THERMOCHEMISTRY

In the field of chemical and metallurgical processes chemical reactions occur with or without mechanical work involvement. **Chemical reactions are nothing but making or breaking of chemical bonds** (in some both may occur). Energy exists in the form of these bonds (in terms of bond strength) and hence energy, most often heat, is required to break these bonds and conversely it is released if bonds are formed. Hence in chemical processes heat may be released or absorbed depending upon whether chemical bonds are formed or broken. If both of these, making and breaking, occur together, then the net algebraic sum of the bond energies shall be released or absorbed, as the case may be. In a heterogeneous process in which gaseous phase is involved the work term may be substantial. Whereas in liquid-liquid or liquid-solid or solid-solid heterogeneous systems the volume change, even though finite, may be negligibly small to contribute any worthwhile work factor in the process. In systems containing a gas phase, if a perceptible volume change occurs in the process the work term may be dominant.

Besides the direct chemical reactions, energy exchanges are also associated with phase changes that may occur in the process, *e.g.,* solid to liquid or liquid to gas or *vice-versa*, in any of the phases involved in the process. Phase transformations in solid state like order-disorder or austenite-martensite transformations are also accompanied by heat exchanges but to a small extent. The process of formation of solutions or de-formation of solutions can contribute a substantial part of heat exchanges occurring in a chemical process.

The field of thermochemistry is to study all the above mentioned heat exchanges during chemical or physico-chemical processes. Calorimeter is often used to study these in a laboratory.

2.9.1 Hess's Law - Corollary of First Law

As the heat exchanged in a process is seen to be a path dependent property, the heat exchanged in chemical or metallurgical processes might also be erroneously viewed as path dependent property.

These processes are studied under either isobaric or isochoric conditions and as shown in Eq. (2.24) and (2.27) as

$$\partial U = \partial q \text{ (for isochoric)} \quad \text{and} \quad dH = \partial q \text{ (for isobaric)}$$

The heat changes in chemical processes are not path dependent but are state functions only. This fact was realised, though not quite in the way as is understood today, by G.H. Hess in 1840. He suggested that **the total heat exchanged in a chemical process is the same irrespective of whether the process occurs in a direct single stage or through multiple stages**. This is now known as **Hess's Law.**

It is now considered as a corollary of first law of thermodynamics. If Hess's law does not hold good then in a cyclic process when the forward process is carried out in say directly one stage and the reverse in multistages then it will result in a net gain of loss of energy. This ought to be zero in a cyclic process and hence both should neutralise each other in such a way that the net amount is zero as per the first law of thermodynamics. Hence, the Hess's law is a corollary of the first law of thermodynamics.

The law can be explained by an illustration such as reduction of iron oxide by carbon or carbon-monoxide gas. The reaction can be carried out in one stage as

$$3Fe_2O_3 + 3C = 2Fe + 3CO \quad \text{...(2.43)}$$

or alternatively in stages as

$$3Fe_2O_3 + C = 2Fe_3O_4 + CO \quad \text{...(2.44)}$$

$$2Fe_3O_4 + 2C = 6FeO + 2CO \quad \text{...(2.45)}$$

and

$$6FeO + 6C = 6Fe + 6CO \quad \text{...(2.46)}$$

The total enthalpy change in both of these routes is exactly the same. This can be proved thus :

The enthalpy change in reaction (2.43), *i.e.*, single stage is given by

$$\Delta H_{(2.43)} = 2H_F + 3H_{CO} - 3H_C - H_{Fe_2O_3} \quad \text{...(2.47)}$$

and enthalpy change in reactions (2.44) to (2.46) are

$$\Delta H_{(2.44)} = 2H_{Fe_3O_4} + H_{CO} - 3H_{Fe_2O_3} - H_C \quad \text{...(2.48)}$$

$$\Delta H_{(2.45)} = 6H_{FeO} + 2H_{CO} - 2H_{Fe_3O_4} - 2H_C \quad \text{...(2.49)}$$

$$\Delta H_{(2.46)} = 6H_{Fe} + 6H_{CO} - 6H_{FeO} - 6H_C \quad \text{...(2.50)}$$

which on addition gives the same equation as in Eq. (2.47).

■ **Example :** Calculate the enthalpy of formation of WC from the following data at 25°C :

(1) $WC + 2\frac{1}{2} O_2 = WO_3 + CO_2$

$\Delta H^\circ = -285800$ cal/g mole

(2) $C + O_2 = CO_2$

$\Delta H^\circ = -940950$ cal/g mole

(3) $W + 1\frac{1}{2} O_2 = WO_3$

$\Delta H^\circ = -195700$ cal/g mole

Solution : WC is formed from C and W as

$$W + C = WC$$

and is obtained by addition of equations 2 and 3 and subtracting from this, equation 1 and the algebraic summation of the $\Delta H°$ values give the required enthalpy change as

$$\Delta H° = \Delta H_2 + \Delta H_3 - \Delta H_1$$

$$= -94050 - 194700 - (-285800)$$

$$= -2950 \text{ cal/g mole}$$

2.10 HEAT CAPACITY CALCULATIONS

Heat capacity for any substance is not a fixed value. For the same material it changes with temperature and the way it changes with temperature varies considerably from substance to substance. The actual relationship has to be established experimentally. However, certain empirical rules have been suggested and are used for approximate calculations. For mono-atomic gases at high temperatures, the specific heat C_V or C_P approaches a value of 3/2R. For diatomic gases the same is close to 5/2R and for solid materials Dulong and Petit suggested that the molal heat capacity is equal to 3R, *i.e.*, $3 \times 1.987 = 5.96$ cal/degree. This statement is known as **Dulong and Petit' Law** (1819). Einstein developed the necessary theory for calculation of heat capacity based on quantum theory with certain assumptions regarding the nature of crystal structure. This is known as **'Einstein's Theory of Specific Heat of Solids'**. This was found to be reasonably valid for high temperature ranges but it failed to explain the observed values of specific heats at low temperatures. This has thus been modified by Debye, that below a certain temperature, known as '**Debye's characteristic temperature**' and, which varies from substance to substance, the specific heat is given by

$$C_P \text{ or } C_V = 234R(T/\theta_D)^3 = 464.5\,(T/\theta_D)^3 \quad \text{...(2.51)}$$

or

$$C_P \text{ or } C_V \propto T^3 \quad \text{...(2.52)}$$

where θ_D is the Debye's characteristic temperature. This is also known as Debye's **T^3 Law** for low temperature heat capacities. The observed values of specific heats of various materials at low temperaures show good agreement with the above equation.

Since the variation of specific heat of any substance with temperature is a complex relationship, after exhaustive experimentation empirical relationship have been developed for a number of elements and compounds in the form

$$C_P = a + bT + cT^{-2} \quad \text{...(2.53)}$$

and

$$C_V = a' + b'T + c'T^{-2} \quad \text{...(2.54)}$$

where a, b, c, a′, b′, and c′ are all constants characteristic of the substance over a certain temperature range. If data are not sufficient the final term may be omitted. Simplification is also possible by using mean specific heat over a certain temperature range as defined as

$$C_P \text{ or } C_V \text{ (mean values)} = \frac{\int_{T_1}^{T_2} C_V \text{ (or } C_P)}{(T_1 - T_2)} \quad \text{...(2.55)}$$

Often tables are available giving the values of constants a, b anc c for a large number of substances.

2.10.1 Kopp Rule

Where specific heat values of compounds are not known from experimentation it can be approximately evaluated by the use of **'Kopp Rule'** suggested in 1864. It says that the specific heat of a compound B shall approximately be given by a formula as

$$\text{Molar sp. Heat of } A_x B_y = x \text{ . sp. heat of A} + y \text{ . sp. heat of B per g atom} \quad \text{...(2.56)}$$

This is fairly reliable for ionic compounds wherein the individual elements obey Dulong and Petit's law.

This is extended to solutions, metallic as well as non-metallic, to work out approximate values of specific heats from the specific heat values of individual components. The specific heat of solutions is equal to the weighted sum of the heat capacities of the components. This can be explained by two examples as :

■ **Example 2.1** A certain slag contains 40% CaO, 35% SiO_2 and 25% Al_2O_3

If the specific heats given are :

CaO – 0.01031 cal/g/0°/mole

SiO_2 – 0.01498 cal/g/0°/mole

Al_2O_3 – 0.01906 cal/g/0°/mole

then the weighted sum of these values shall be

$$= 0.01031 \times 0.40 + 0.01498 \times 0.35 + 0.01906 \times 0.25$$

= 0.0095 cal/g/0°/mole for the slag under consideration.

■ **Example 2.2** If the sp. heat of copper is 0.096671 and that of silver is 0.05582 cal/g/0°, what will be the value of sp. heat of an alloy of 60% Cu and 40% Ag ?

The weighted sum of shall be

$$= 0.096671 \times 0.6 + 0.05582 \times 0.4$$

= 0.077322 cal/g/0°

2.10.2 Variation of Enthalpy with Temperature

The heat content or enthalpy required to heat a substance over a certain temperature range or that which would be evolved on cooling is of direct relevance in chemical or metallurgical processes. The basic equation of enthalpy as it varies with temperature is

$$(\Delta H)_P = \int_{T_1}^{T_2} C_P \, dT$$

A relationship as shown in Fig. 2.2 can be drawn for ΔH as it varies with temperature. The enthalpy change in heating the material from T_1 to T_2 temperatures shall be equal to the area under the curve between T_1 and T_2. A varieties of such relationships would be observed for different substances depending upon their nature.

$$\Delta H = H_2 - H_1 = \int_{T_1}^{T_2} (a + b \, . \, T + c \, . \, T^{-2}) \, dT$$

$$= (T_2 - T_1)\, a + (T_2 - T_1)\, bT^2 + c\left(\frac{1}{T_2} - \frac{1}{T_1}\right) \quad \text{...(2.57)}$$

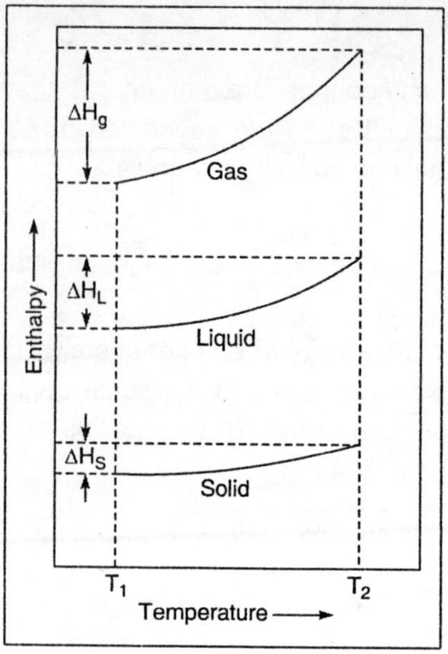

Fig. 2.2 Schematic variation of enthalpy of solid, liquid and gas phases for different materials.

Knowing the values of T_1 and T_2 and a, b and c as constants related to C_P the enthalpy change can be calculated. Since there is nothing like absolute enthalpy a relative value is only calculated. It is therefore customary to assign H_{298}°K as the base level for comparison and actual values of enthalpy changes as

$$H_{298}\,^{\circ}K - H_{298}\,^{\circ}K \qquad ...(2.58)$$

are estimated from the specific heat data and are compiled in tabular form for ready reference for various pure elements and compounds in the form of H_{298} °K data.

Normally reactions are carried out with the reactants being at room temperature to start with. The reactants may be heated to some elevated temperature and then the reaction occurs to produce the products.

For a reaction say $A + B = AB$

the enthalpy change at 298°K can be evaluated knowing the value of ΔH_{298} °K values of reactants and products as

$$\Delta H_{298}^{\circ} = \Delta H_{298}^{\circ}(AB) - \Delta H_{298}^{\circ}(A) - \Delta H_{298}^{\circ}(B) \qquad ...(2.59)$$

In a generalised way the value of ΔH_{298} °K can be evaluated from the knowledge of ΔH_{298} °K and values of specific heats of reactants and products involved in the reaction knowing the Kirchoff's law as explained in the following section.

■ **Example 2.3** For metallothermic reduction of solid oxides to be a feasible proposition the enthalpy change required is around 600 cal/g weight of the system. Establish which one of the following reaction is a viable proposition?

1. $3PbO + 2Al = Al_2O_3 + 3Pb$
2. $2PbO + Si = SiO_2 + 2Pb$
3. $Fe_2O_3 + 2Al = 2Fe + Al_2O_3$

4. $3ZnO + 2Al = 3Zn + Al_2O_3$

5. $2MnO + Si = SiO_2 + 2Mn$

Given for the solid reactants and products :

$2PbO + O_2 = 2PbO$ $\Delta H^\circ = -104500$ cal/g/mole

$4/3Al + O_2 = 2/3Al_2O_3$ $\Delta H^\circ = -256600$ cal/g/mole

$4/3Fe + O_2 = 2/3Fe_2O_3$ $\Delta H^\circ = -196000$ cal/g/mole

$2Zn + O_2 = 2ZnO$ $\Delta H^\circ = -83000$ cal/g/mole

$2Mn + O_2 = 2MnO$ $\Delta H^\circ = -92000$ cal/g/mole

$Si + O_2 = SiO_2$ $\Delta H^\circ = -209000$ cal/g/mole

therefore

$$2PbO + 4/3Al = 2/3Al_2O_2 + 2Pb$$

from Hess's law

$$\Delta H^\circ = -152100$$

$$\Delta H^\circ = -152100$$

$$3PbO + 2Al = Al_2O_3 + 3Pb$$

$$\Delta H^\circ = -228150$$

$$\text{sp. heat/g of system} = -\frac{228150}{723} = -315 \text{ cal}$$

where 723 is the sum total of gram atoms of products or reactants.

Therefore, even if pure oxide was reduced it would not be possible to carry this reduction. In fact in practice the oxide will be impure, that is additional dead weight will be there, and the specific heat will still further decrease and make it non-practical.

In the same way the others can be calculated and assessed to find which one is a feasible reaction. It will be observed that the reduction of iron oxide by aluminium will be a very favourable one and that is the **thermal process** of welding.

■ **Example 2.4** Calculate the temperature rise when 0.01% P dissolved in molten iron is oxidised by supply of oxygen and solid lime at 27°C during a steel making process to make liquid calcium phosphate slag at 1750°K by a reaction :

$$4[P]_{1750^\circ} + 5\{O_2\}_{300^\circ} + 6 < CaO >_{300^\circ} = 2(3CaO.P_2O_5)_{1750^\circ}$$

Given :

1. $\{P_2\} = 2[P]$ $\Delta H_1 = -58500$ cal
2. $4/5\{P_2\} + \{O_2\} = 2/5 <P_2O_5>$ $\Delta H_2 = -15160$ cal
3. $\{P_2O_5\} = < P_2O_5 >$ $\Delta H_3 = -3600$ cal
4. $6/5 <CaO> + 2/5 \{P_2O_5\} = 2/5 <CaO . P_2O_5>$ $\Delta H_4 = -405300$ cal
5. mean sp. heat of oxygen gas, solid CaO and molten iron are 0.255, 0.1 and 0.1 cal/degree/g respectively.
6. Heat of fusion of phosphate = + 4000 cal/mole

Solution : The process is to be carried out as shown Fig. 2.3.

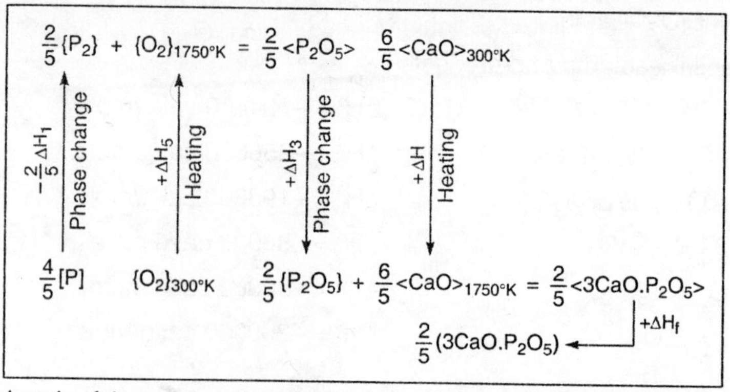

Fig. 2.3 The physical mode of the reaction of oxidation of phosphorus from molten iron to give molten phosphate.

The final reaction can be built up in stages using Hess's law as follows to obtain the overall enthalpy change :

$2[P] = \{P_2\}$ $\qquad \Delta H_1 = +58500$ cals

$4/5[P] = 2/5\{P_2\}$ $\qquad \Delta H_1 = +58500 \times 2/5 = 23400$

$2/5\{P_2\} + \{O_2\} = 2/5 <P_2O_5>$ $\qquad \Delta H_2 = -15160$ cals

$4/5[P] + \{O_2\}_{300^\circ} = 2/5 <P_2O_5>$ $\qquad \Delta H = \Delta H_1 + \Delta H_2 = 8340$ cals

and to this ΔH of the reaction 4 and add solid CaO to it to give

$$4.5[P] + \{O_2\} + 2/5 <CaO> = 2/5 <CaO.P_2O_5>$$

$$\Delta H = 8340 - 405300 \times 2/5 = -153380 \text{ cals}$$

Algebraically account for heating of reactants oxygen gas and solid lime to 1750°K and for melting of calcium phosphate.

These are :

$\{O_2\}_{300^\circ} = \{O_2\}_{1750^\circ}$ $\qquad \Delta H = C_P \times dT \times \text{mass}$

$= 0.255 \times 1450 \times 32$

$= 11840$ cal

$6/5 <CaO>_{300^\circ} = 6/5 <CaO>_{1750^\circ}$ $\qquad \Delta H = 0.1 \times 1450 \times 56 \times 6/5$

$= +9744$ cals

and $\quad 2/5 <CaO . P_2O_5> = 2/5 (3CaO . P_2O_5)$ $\qquad \Delta H = 4000 \times 2/5$

$= 1600$ cals

All these will have to be met from the exothermic heat to obtain the liquid product phosphate. The net enthalpy change is therefore

$$= -153380 + 11840 + 9744 + 1600 = -130636 \text{ cals}$$

This enthalpy is evolved when (4/5) × 31 = 24.8 g of P is oxidised in molten iron to get equivalent molten phosphate. When 100 g of melt is taken and 0.1 g of P is oxidised adiabatically the temperature rise shall be

$$dT = \frac{\Delta H}{C_P \cdot \text{mass}} = \frac{1306}{0.1 \times 100 \times 4/5 \times 31} = 5.2^\circ C$$

2.10.3 Enthalpy Change during Phase Change

When phase changes like solid to liquid or liquid to gas or *vice-versa* or order-disorder changes or solid state phase changes like austenite to ferrite and/or cementite occur, energy exchange is a must and hence the enthalpy changes in such processes are given by the corresponding latent heats of such processes.

During heating of a material (or even cooling in reverse) from say T_1 to T_2 temperature if a phase change occurs in between then the enthalpy change will be given by

Process enthalpy change

$$\Delta H^{\circ}_{(T_1 - T_2)} = \int_{T_1}^{T(f)} C_P\, dT \text{ (phase 1)} + \Delta H_f^{\circ} + \int_{T_1}^{T_2} C_P\, dT \text{ (phase 2)} \qquad ...(2.60)$$

where ΔH_f° is the enthalpy of the phase change taking place at temperature Tf which is in between the temperatures T_1 and T_2. This equation shall get extended if more then one such phase changes occur in the over all process by the addition of similar terms in between the temperatures T_1 and T_2. For example if manganese is heated from 500°C to say 1250°C, the overall enthalpy change shall include the enthalpy changes accompanying the phase changes for the intermediate phase changes such as Mn(α) to Mn(β) at 720°C, Mn(β) to Mn(γ) at 1100°C and Mn(γ) to Mn(δ) at 1136°C and therefore the enthalpy equation shall take the form as below :

$$\Delta H = \int_{773\text{-}993°K} C_P(Mn_\alpha)\, dT + \Delta H_{993°K}(\alpha - \beta) + \int_{993\text{-}1373°K} C_P(Mn_\beta)\, dT + \Delta H_{1373°K}(\beta - \gamma) +$$

$$\int_{1373\text{-}1409°K} C_P(Mn_\gamma)\, dT + \Delta H_{1409°K}(\gamma - \delta) + \int_{1409\text{-}1523°K} C_P(Mn_\delta)\, dT \qquad ...(2.61)$$

This evaluation of enthalpy as shown mathematically above can also be depicted schematically as shown in Fig. 2.4. It would be clear from both that this evaluation of enthalpy

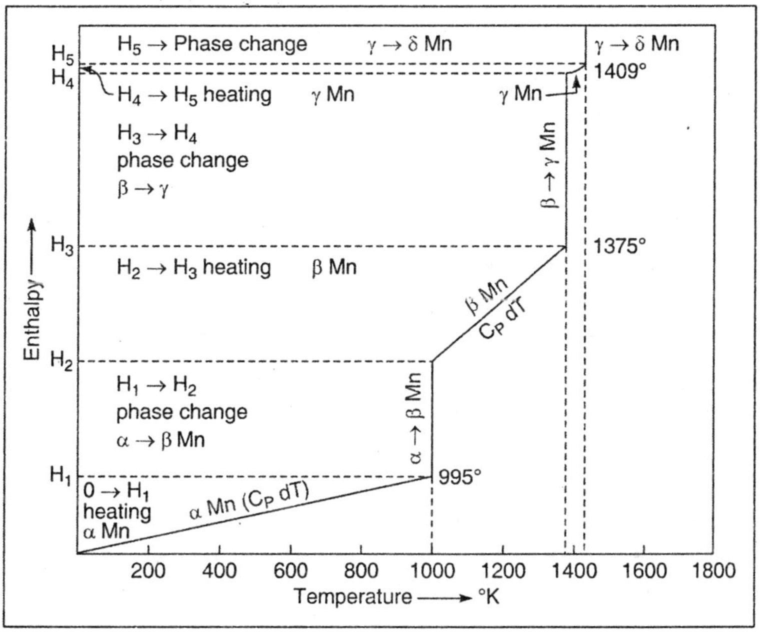

Fig. 2.4 Representation of enthalpy change on heating manganese over a temperature range involving phase changes in manganese.

necessarily requires the knowledge of C_P values of all such substances (crystal forms) involved in the overall process. It is also clear that if enthalpy is known at some temperature then its value at some higher temperature can be found out as

$$\Delta H_{T_2} = H_{T_1} + \int_{T_1}^{T_2} C_P \, dT \qquad \text{...(2.62)}$$

and with appropriate changes in sign in this equation it can also be determined at a lower temperature as well. It means knowing the value at some temperature on the graph in Fig. 2.2 the remaining can be added to or subtracted from it.

2.11 KIRCHOFF'S LAW

This is the law which helps to evaluate enthalpy of a reaction at a temperature when it is known at some other temperature. The variation of enthalpies of reactants and products with temperature for a certain reaction are shown in Fig. 2.5. Initially the reactants are at state A at temperature T_i (initial) and the final products are at state F at temperature T_f (final). The reaction can be carried out by the route ABCDF or ABEDF wherein the reaction occurs at temperature T_1 and T_2 respectively. In either case the enthalpy change is the same.

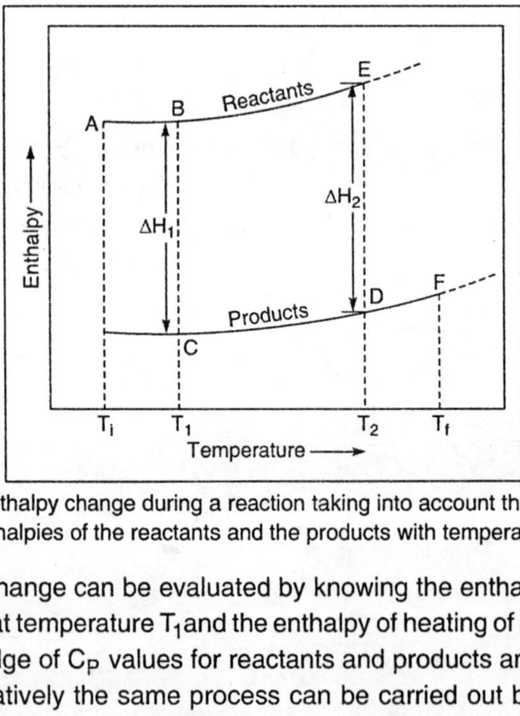

Fig. 2.5 Enthalpy change during a reaction taking into account the changes in enthalpies of the reactants and the products with temperature.

The total enthalpy change can be evaluated by knowing the enthalpy of heating from A to B, then enthalpy of reaction at temperature T_1 and the enthalpy of heating of products to temperature T_f. This requires the knowledge of C_P values for reactants and products and also enthalpy change of the reaction at T_1. Alternatively the same process can be carried out by heating the reactants to temperature T_2 followed by reaction at temperature T_2 and then heating of the products to T_f. This also needs the knowledge of C_P values for reactants and products and the enthalpy change of the reaction at temperature T_2. In either case the overall enthalpy change is the same since the initial and the final states are the same in both the cases. For a cyclic process BCDEB the net enthalpy change is zero and therefore

$$\Delta H_{T_1} + \int_{T_1}^{T_2} C_P\,(\text{prod})\,dT - \Delta H_{T_2} + \int_{T_1}^{T_2} C_P\,(\text{react})\,dT = 0$$

B → C C → D D → E E → B B → B

or
$$\Delta H_{T_2} - \Delta H_{T_1} = \int_{T_1}^{T_2} [C_P\,(\text{prod}) - C_P\,(\text{react})]\,dT \qquad ...(2.63)$$

if it is shown as
$$[C_P\,(\text{prod}) - C_P\,(\text{react})] = \Delta C_P$$

$$\Delta H_{T_2} - \Delta H_{T_1} = \int_{T_1}^{T_2} \Delta C_P\,dT \qquad ...(2.64)$$

Knowing the values of specific heats for reactants and products and enthalpy change at some temperature (T_1) the enthalpy change of some other temperature (T_2) can be calculated from the above equation. Thus if enthalpy value is known at either T_1 or T_2 the other can be calculated from the above equation. Alternatively the change in the enthalpy values between temperature T_1 and T_2 can be calculated as

$$\Delta H_{T_2} - \Delta H_{T_1} = \partial \Delta H = \int_{T_1}^{T_2} \Delta C_P\,dT$$

or
$$\left(\frac{\partial \Delta H}{dT}\right)_P = \Delta C_P \qquad ...(2.65)$$

The subscript **P** is to indicate that this is valid under conditions of constant pressure as per the original equation used in the derivation. This is the statement of **Kirchoff's law. It means that the rate of change of enthalpy of a process or a reaction with temperature is given by the difference of the specific heats at constant pressure of products and the reactants.** The same can be arrived at as

$$C_P\,(\text{prod}) = \left(\frac{\partial H}{\partial T}\right)_P \quad \text{and} \quad C_P\,(\text{react}) = \left(\frac{\partial H}{\partial T}\right)_P$$

by subtracting
$$\frac{\partial H_{prod} - \partial H_{react}}{\partial T} = C_{P(prod)} - C_{P(react)}$$

and since the left hand side is the ΔH of the reaction

$$\left(\frac{\partial \Delta H}{dT}\right)_P = \Delta C_P \quad \text{as shown in Eq. (2.65)}$$

If the reaction is carried out under adiabatic conditions either by route ABCDF and ABEDF the total enthalpy change will be given by either

$$\int_A^B C_P\,dT + \Delta H_{T_1} + \int_C^F C_P\,dT$$

reactants reaction products

or
$$\int_A^E C_P\,dT + \Delta H_T + \int_D^F C_P\,dT$$

reactants reaction products

The temperature rise of the product can thus be calculated from these working the mass of the reactants and the products. Basically the reaction process is either ABCDE or ABEDF and this includes the heating and reaction steps. Given the required details of any such process the remaining unknown can be found out.

■ **Example 2.5** Calculate the enthalpy of vaporisation of Mn at 1900°K (sub-boiling point temp.) and of fusion at 1200°K (sub-melting point) when these values are known at their normal boiling and melting points and when the specific heat values of solid, liquid and gaseous Mn are known.

Solution : The process has to be imagined to occur as shown in Fig. 2.6. The overall enthalpy change will be

$$= + \int_{1200}^{1580} C_P \, dT\,(\text{solid}) + \Delta H_f \;(1500^\circ \text{K}) + \int_{1580}^{1200} C_P \, dT(\text{liq}) - \Delta H_f \;(1200^\circ \text{K})$$

which is equal to zero for a cyclic process.

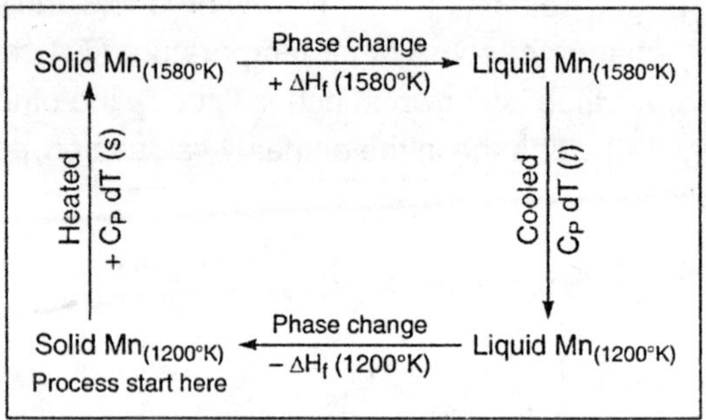

Fig. 2.6 Imaginary scheme of carrying out the process to obtain the desired end product.

Hence

$$\Delta H_f \;(1200^\circ \text{K}) = \int_{1200}^{1580} C_P \, dT(\text{solid}) + \Delta H_f \;(1580^\circ \text{K}) + \int_{1200}^{1580} C_P \, dT\;(\text{liq})$$

All the values on the right side are known as given and hence ΔH_f (1200° K) can be calculated. Similar exercise carried out for vaporisation can give the value of ΔH_V (1900° K).

2.12 ADIABATIC FLAME TEMPERATURE

It is often advisable to know the maximum temperature of the flame that can be achieved theoretically when fuel is burned. This is possible to be calculated assuming adiabatic conditions of burning. It means that whatever heat is liberated in the combustion of whatever constituents (C, H, etc.) that are present shall be available to raise the temperature of the products that make the flame. It also requires that the C_P values for CO, CO_2, H_2O, etc. be known for that temperature range. If it is assumed that ΔH is the enthalpy of the overall process

$$\text{Reactants at } T_1 \rightarrow \text{burning} \rightarrow \text{Products at } T_2$$

and if
$$C + O_2 = CO_2$$
and
$$2H + O_2 = H_2O$$

are the combustion reactions then the products can be ascertained from the composition of the fuel and amount of air and or oxygen supplied for its burning. And

$$\Delta H = \int_{298^\circ K}^{T_1} [C_P\,(CO_2) \,.\, n_{CO_2} + C_P\,(H_2O) \,.\, n_{H_2O} + C_P\,(O_2) \,.\, n_{O_2} + C_P\,(N_2) \,.\, n_{N_2}]\, dT$$

The excess oxygen term will appear only if excess oxygen or air is supplied and similarly nitrogen term will appear only if air is supplied for combustion and in proportion to 79 : 21. The above will have to be integrated to obtain the unknown T as the adiabatic flame temperature.

2.13 CONCLUDING REMARKS

1. Energy cannot be created nor can it be destroyed, it changes only its form.
2. Energy can be interconverted.
3. Although the first law as such does not impose any limitation on such interconversion and, as is observed in practice, that mechanical work can be fully converted to heat energy, yet out of all conceivable changes, only some actually occur in practice in fully converting the form, whereas many others do not do so.
4. There is a certain tendency for the conversion of energy to occur in some preferential direction rather than for either direction to be equally possible. This limitation shall be dealt with in the following chapter.

This anomaly, though not conceived in the first law of thermodynamics, can be explained only in terms of the second law of thermodynamics. This will be discussed in the following chapter.

□□□

Chapter 3

Second Law of Thermodynamics

Entropy and its Derivatives

3.1 INTRODUCTION

A thermodynamic system in equilibrium shall stay in equilibrium unless acted upon by external constraints to make it move away from the equilibrium position. However, if the system is away from equilibrium it will tend to achieve the equilibrium state. Systems, which are away from equilibrium, upon initiation, shall move towards equilibrium and such processes are referred to as **'natural'** or **'spontaneous'** or **'irreversible'** processes. Examples are heat flow from a hotter to a colder body (in contact with) or depressurisation of an inflated tube in a low pressure surroundings or a free fall under gravity and so on. The spontaneous change from an existing state to an equilibrium state is possible because in the existing state the system happens to be at higher potential and which is the driving force for the change to occur. Higher temperature is therefore a higher thermal potential and which makes heat flow from higher to a lower temperature. Similarly, higher pressure is a higher mechanical potential and so on.

If a system is in equilibrium and if it is to be moved in the reverse direction in the above examples, it would be termed as **'unnatural'** or **non-spontaneous** processes. Therefore, water cannot be raised to an overhead tank unless energy, in the form of motor-pump set is provided to the system from the surroundings. Similarly, heat will not flow from a colder to a hotter body unless aided by the surroundings in the form of compressor energy as in a refrigerator. These are the examples that can be observed but in chemical processes it is not readily possible to assess as to what is a natural process or an unnatural process. This can only be evaluated from the knowledge of equilibrium state of the system.

The knowledge of equilibrium CO/CO_2 ratio in contact with Fe or FeO can only guide us as to how to prevent oxidation of iron or effect reduction of iron oxide by providing a suitable CO/CO_2 gas mixture as surroundings. The first law of thermodynamics merely states the inter-conversion of heat and work but not the direction in which this will occur. It also does not give any indication of internal energy being a potential to make a process take place. The second law is able to devise a potential term capable of driving the process in the form of a force.

The first law is based on the essential observation that mechanical energy is fully converted to heat (from Joule's adiabatic experiment). It is possible to carry out this process in the reverse direction, *i.e.*, heat to be converted fully to mechanical energy? This was of direct relevance to understand the working of steam engines in the early days and also to improve upon their efficiencies. These efforts to appreciate the performance of steam engines in the 18th and 19th century ultimately led to the formulation of second law of thermodynamics.

It would be worthwhile to study an ideal gas engine producing work by taking heat from a reservoir. Fig. 3.1 shows the P-V relationship of such an engine. It takes heat from the surrounding and expands isothermally from state A to state B doing mechanical work equal to $\int_A^B PV$. In order to run it as an engine this must be done repetitively to produce work by taking heat from the surroundings. The system therefore must return back to A, its original state. This is not possible in one single adiabatic compression. At least two reversible processes, one of which may be isothermal, are needed to revert the gas system to its original state A. These are shown in Fig. 3.1 as processes BC and CA. Both of these will have to be on the lower side of the isothermal curve AB. The work $\int_B^C PV$ and $\int_C^A PV$ together must be and is less than $\int_A^B PV$ to leave out net mechanical work done by the system in a cyclic process.

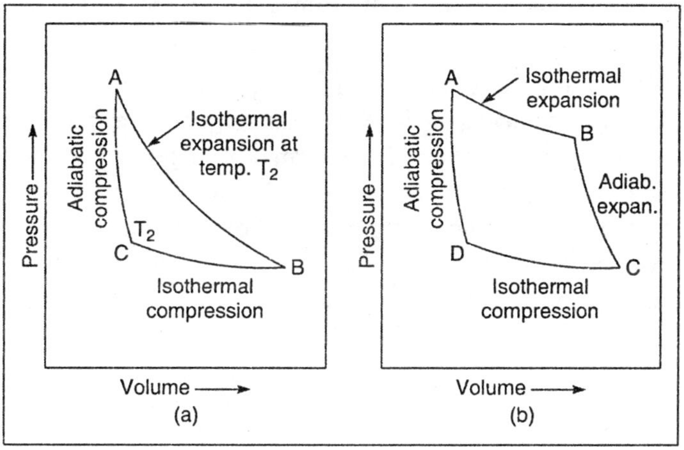

Fig. 3.1 Typical P-V diagrams in cyclic processes related to a gas engine : (a) shows a three stage cycle and (b) shows a four stage cycle.

In such a cyclic process involving minimum three different processes the gas takes heat at a temperature T_1 from the surrounding and expands to do work. On reverse side it has to isothermally compress first, desorbing the equivalent heat to the surrounding, and then to adiabatically compress it further to attain the original state A. This necessitates that the system must give out heat during the isothermal compression at a temperature lower than at which it earlier accepted it for doing the work by expansion. The adiabatic compression shall raise the temperature to T_1 again when the original state A is achieved. This whole process, for conversion of heat to work necessarily produces a permanent change in the second surrounding reservoir by way of release of heat into it. This inevitable transfer of heat to the surrounding at a lower temperature, which is a permanent change in the second surrounding, to carry out the conversion of heat to work is called by some as **'compensation'**. Hence, heat can not be converted to work without such compensation. It also means that the entire heat that is absorbed initially is not converted into useful work but a part of it is rejected to the second reservoir as of necessity. In other words **'in a cyclic process it is just not possible to convert all heat into mechanical work, although it is possible to convert all work**

into heat'. In a non-cyclic process it may be possible to convert all heat into work but in a cyclic process it is not so. The important consideration is therefore that in a cyclic process all heat can not be converted into work fully.

If heat is to flow from a relatively colder to a hotter body, it is only possible by attaching the colder body to a gas engine which will receive heat from the colder body by way of its expansion and by adiabatic compression raise its temperature to that higher than the hotter body and thus desorb heat taken from colder body to a hotter body. The work done by he gas engine as surrounding is the required compensation.

The second law assumed the form in the early days in the words of Carnot as **'it is not possible to construct a cyclic device to convert heat into work fully on its own'**. The same thing was put by Kelvin as **'it is just not possible to convert heat into work in a cyclic process without compensation'**. A later version of second law by Clausius states that, **'in a cyclic process it is impossible to transfer heat from a colder to a hotter body without producing permanent changes or compensation in the surrounding'**. Max Plank suggested that **'it is impossible to construct a cyclic device extracting heat from a reservoir and produce thermal effects elsewhere'** (compensation).

The first law states that a system is a energy reservoir and hence can accept energy in any form from the surroundings and give it out in the same or other energy forms in the same or different amount. In the above example the system (in the form of ideal gas engine and piston) accepts heat from surroundings but does not convert it all to mechanical work. A part of this energy is converted to work and a part is parted off to another system at a lower temperature. In other words in a closed cycle operation converting heat into work a part of the heat input has to be rejected to a colder reservoir. The efficiency of this engine is given by

$$\frac{\text{network done on to the surrounding}}{\text{heat supplied to the system}}$$

3.2 CARNOT CYCLE

In order to work out the efficiency of a steam engine Carnot in 1824 imagined a cyclic operation consisting of two isothermal and two adiabatic, each with a expansion and a compression, thus totalling four stages in a certain way. This is shown in Fig. 3.2 and is known as **'Carnot Cycle'**. The operation of such an idealised gas engine can be imagined to start from any point. Let us say it starts from A. The engine with ideal gas as the working substance is initially at A at pressure P_1, volume V_1 and temperature T_1 conditions. It is put in contact with a thermostat at temperature T_1 and is allowed to expand, taking heat q_1 and doing work w_1 on the surrounding. It thus reaches B at corresponding state properties as P_2, V_2 and T_1 remains the same, at the end of the isothermal process of expansion. Now the thermostat is removed and it is insulated and allowed to expand further adiabatically to conditions P_3, V_3 and T_2 doing further work w_2 on the surrounding and thereby reaching position C. Now T_2 has to be less than T_1 because of loss of adiabatic work from the system to the surrounding. It is now again put in contact with a thermostat at temperature T_2 and isothermally compressed putting in work w_2 to achieve conditions P_4, V_4 and T_2 at position D. In this it gives out heat q_2 to the reservoir because of work w_3 being put into it from the surrounding for compressing it from position C to D. The thermostat is now removed at D and system is insulated

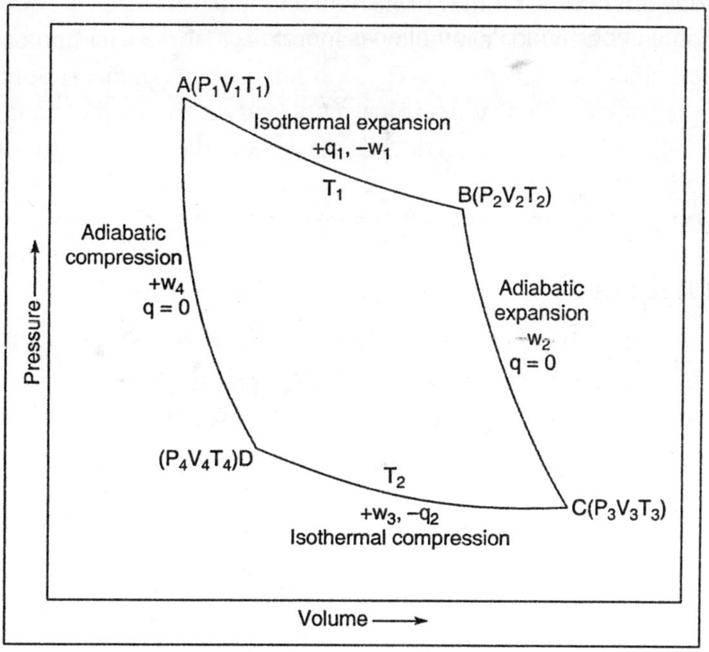

Fig. 3.2 Typical Carnot cycle operation.

and further compressed putting w_4 into the system from the surrounding which would raise the temperature from T_2 to T_1 and pressure and volume to the original P_1 and V_1 and thereby completing the cycle. On the whole the situation is as follows :

The gas engine takes heat q_1 at T_1 temperature and does work w_1 and rejects heat q_2 at temperature T_2 while receiving work w_2. The two adiabatic processes, one of expansion and the other of compression nullify each other as far as the net change is concerned. Also since it is closed cycle the net work done and the net heat absorbed are equal and hence net energy exchange is zero.

Therefore, the total work done can be equated to the total heat exchanged as follows :

Isothermal work done during V_1 to V_2 and during V_3 to V_4 equals the heat exchanged during the same period as q_1 and q_2. Hence as per Eq. (2.33) in Section (2.8.1) the two isothermal work terms can be written as

$$RT_1 \ln (V_2/V_1) + RT_2 \ln (V_4/V_3) = q_1 + q_2 \quad ...(3.1)$$

In order to obtain maximum work all operations are carried out reversibly.

From first law putting $q = 0$ for an adiabatic process

gives $dU = -w = -P\,dV$

and since $dU = C_V\,dT$ and $C_V\,dT = -P\,dV$

for an ideal gas $C_V\,dT = -(RT/V)\,dV$

and rearranging the terms it becomes

$$\frac{dV}{V} = -\frac{C_V}{R} \cdot \frac{dT}{T}$$

integrating between the limits T_1 and T_2 when V_2 expands to V_3

$$\ln (V_2/V_3) = -\frac{C_V}{R} . \ln (T_2/T_1) \quad ...(3.2)$$

$$\frac{V_2}{V_3} = \left(\frac{T_2}{T_1}\right)^{-C_V/R} = \frac{V_1}{V_4}$$

hence on rearranging

$$\frac{V_2}{V_1} = \frac{V_3}{V_4} \quad ...(3.3)$$

Hence Eq. (3.1) reduces to

$$RT_1 \ln (V_2/V_3) - RT_2 \ln (V_2/V_1) = q_1 + q_2 \quad ...(3.4)$$

or it means

$$R(T_1 - T_2) \ln (V_2/V_1) = q_1 + q_2$$

and since

$$RT_1 \ln (V_2/V_1) = q_1 \quad \text{or} \quad q_1/T_1 = R \ln (V_2/V_1)$$

therefore,

$$(T_1 - T_2)(q_1/T_1) = q_1 + q_2 \quad ...(3.5)$$

or

$$\frac{T_1 - T_2}{T_1} = \frac{q_1 - q_2}{q_1} \quad ...(3.6)$$

that is

$$1 - (T_2/T_1) = 1 + (q_1/q_1)$$

or

$$\frac{q_2}{T_2} + \frac{q_1}{T_1} = 0 \quad ...(3.7)$$

in other words $\phi = \frac{q_1(rev)}{T_1} = 0$ as all the processes in Carnot cycle are carried out reversibly.

The cyclic summation of q_{rev}/T being zero holds true for any substance in the Carnot cycle engine and also for its any state like solid, liquid or gas as there is no condition-precedent like this in arriving at the above relationship. Hence the property q_{rev}/T is a fundamental property and being zero in a cyclic process, it must be a state function and not a path function. The q_{rev}, otherwise referred to as path function, denotes here only the actual amount of heat exchanged and the temperature at which it is exchanged. As was observed earlier while deducing the relationships for C_V and dU and, C_P and dH that, the term q/dT is also equivalent to dU under constant volume conditions and to dH under constant pressure condition and thereby become state functions, so also here, the term q_{rev}/T is a state function and is designated by the term **'entropy'** and often denoted by the letter **'S'** and hence,

$$\phi dS = 0 \quad \text{or} \quad dS = \frac{q(rev)}{T} \quad \text{or} \quad \Delta S = \frac{q(rev)}{dT} \quad ...(3.8)$$

The equation (3.6) in fact means the thermodynamic efficiency of the Carnot engine in the instant case. The total work done by the engine operating as mentioned above is $q_1 + q_2$ and is divided by the heat input into the system as q_1. This factor is nothing but efficiency of the engine. It depends solely on the two temperature limits between which the engine operates. The thermodynamic engine efficiency is therefore independent of the working substance and depends only on the two temperatures. Since T_2 cannot be zero cent per cent conversion of heat into work is just not possible. In other words all heat cannot be fully converted into work. The second law is thus proved. The conversion of heat into work is accompanied by the permanent change in the reservoir at lower temperature wherein a part of the heat supplied by the main high temperature reservoir, is dumped, and which is the required compensation.

3.3 ENTROPY CHANGE IN NATURAL PROCESSES

All natural processes or irreversible processes or those which are away from equilibrium, when initiated they all result in increase in entropy as will be obvious from the following examples :

Let us imagine that heat flows from a higher temperature T_1 reservoir to a lower temperature reservoir T_2 through a conductor under steady state conditions. The conductor be considered as the system in this instant case and the reservoirs as surrounding. Under steady state conditions there will be temperature gradient T_1 to T_2 across the conductor. There is no change in the conductor as a system during the transfer of heat from T_1 reservoir to T_2 reservoir. The conductor receives q/T_1 entropy from reservoir at T_1 and gives out entropy q/T_2 to reservoir at T_2. Since $T_1 > T_2$

$$\frac{q}{T_2} > \frac{q}{T_1}$$

and hence the system receives less entropy and gives out more. Since the flow of heat is a natural process the conclusion is that a **natural process creates or generates entropy.** The reverse process of heat flow from a colder to a hotter body is not possible as per second law of thermodynamics and therefore entropy can not be decreased just like that without compensation. Another examples of an inflated tube can be cited. On puncturing, it gets deflated. The net change is increase of entropy. Also the kinetic energy of a body gets decreased, in a natural process by frictional forces, wherein heat is generated. It means increase of entropy. Similarly, if some weight falls under gravity say in a mixture of ice and water some ice is melted by the potential energy released by the fall and the impact. It means entropy of ice and water has increased. The same argument can thus be put for all natural processes that **entropy increases when natural processes occur.**

It is difficult to obviously prove it in natural chemical processes but it is so.

The Carnot cycle opera' s under reversible conditions. If the cycle is carried out under identical state positions but, under irreversible conditions then the work done in each stage shall be less than that done under reversible conditions. It means that the work done shall not be equal to the heat exchanged but less than heat exchanged and hence for a irreversible natural process involved in a Carnot cycle the two work terms in Eq. (3.1).

$$RT_1 \ln\left(\frac{V_2}{V_1}\right) + RT_2 \ln\left(\frac{V_4}{V_3}\right) < (q_1 + q_2) \qquad \ldots(3.9)$$

On processing these as was done earlier while deriving the Carnot cycle this inequality therefore finally leads to

$$\phi \Sigma \frac{q\,(\text{irrev})}{T} > 0 \qquad \ldots(3.10)$$

for Carnot cycle involving irreversible processes. For any isolated system therefore

$$\Delta S = \text{or} > 0$$

depending upon whether the process is reversible for irreversible (natural).

This equation is the expression of universal law of increase of entropy which states **'the entropy of a system in isolation can never decrease'**. It means the isolated system has to go towards more randomness or chaos.

If the argument is carried further and if the process is to be made to go like melting of ice or lifting of water into a overhead tank or reduction of oxides of baser elements and so on, then the entropy change is less than zero or $dS < 0$. In all the above discussion the entropy change means total entropy of the system and the surroundings put together. These processes are all unnatural and hence it can be deducted that the entropy decreases in unnatural processes.

For any given conditions of an isolated system there is only one and one true equilibrium state of the system and therefore the entropy law can be stated as that **'the conditions of a system in isolation can not be varied in such a way as to decrease the entropy'**.

If a solid is melted into a liquid by supply of heat at the same temperature in the form of latent heat of melting then where is this energy? The temperature, pressure are the same and the volume is practically the same. For sure the minor change in volume accompanying the melting process can not account for the huge amount of latent heat supplied as work done. The internal energy is changed because of supply of heat and one may say that

$$dU = q \text{ (latent head)}$$

but such an equation hardly conveys anything beyond the appreciation of first law of thermodynamics. The Carnot cycle states that the latent heat has resulted in increase in the entropy of the system equivalent to

$$\frac{\text{Latent heat of fusion}}{\text{Melting temperature}} = \frac{H_f}{\text{M.P.}} = \frac{q}{T} = dS \qquad \text{...(3.11)}$$

Physically the rigid atomic arrangement of solid has, on melting, become more flexible with more mobility for atoms/molecules, etc. **Is entropy therefore a measure of this mobility?**

The same argument holds true for liquid to vapour transition at the same temperature. Now the volume change and consequent mechanical work are not equivalent to latent head of vaporisation. This is again a case of increase of entropy as

$$\frac{\text{Latent heat of vaporisation}}{\text{boiling point}} = \frac{H_V}{\text{B.P.}} = \frac{q}{T} = dS \qquad \text{...(3.12)}$$

The randomness of atoms/molecules in vapour is far more than that in liquid. Therefore, **is entropy a measure of randomness again?** The same argument can be put forth for mere heating as well.

Surely entropy is therefore a measure of randomness of atoms or molecules in the system. In solid state, phase changes like crystal changes or order-disorder transformations and the like occur, then the entropy change as a measure of randomness is more apparent.

Since heat supply or heat evolution is an unalienable part of chemical and metallurgical processes, the associated entropy change in the system, as a consequence, is additional and more important parameter for study of energy changes during such processes.

This can best be illustrated from the example of order-disorder change. The two structures one as ordered and the other disordered are shown in Fig. 3.3. These two arrangements must have two

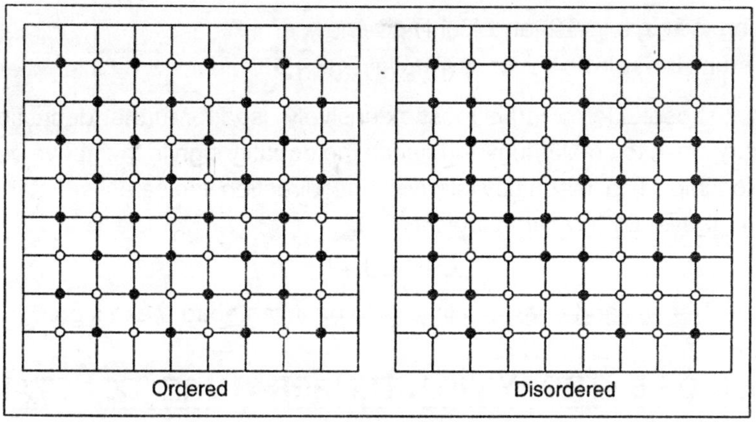

Fig. 3.3 The ordered and the disordered arrangements in a solid crystal lattice. The dark and the hollow circles represent two different types of atoms.

different energies associated with them since the pressure, volume and the temperature are all the same in both the cases. In changing from one form to another some heat say q will have to be exchanged at say a temperature T°K. This is based on the basic fundamental that two different systems with similar chemistry must have two different energies associated with them. If the associated energies were exactly the same then there is no reason for them to be two different systems because P, V and T are all alike. The two therefore differ in only one form and that is the configurational arrangement of the atoms in them. This energy difference can only be described in terms of its entropy alone which is different for different configurations and is described as q/T, *i.e.,* dS, amount of energy.

3.4 COMBINED STATEMENT OF THE FIRST AND THE SECOND LAW

If a system is capable of doing only mechanical work the first law equation can be put as given earlier as Eq. (2.6) as

$$dU = dq - P\,dV$$

and although it is not exact differential equation it is all that first law can do. This equation holds true for a reversible process. The second law for a reversible process gives

$$dS = q/T \quad \text{or} \quad q = T\,dS$$

The two laws therefore can lead to

$$dU = T\,dS - P\,dV \quad \text{or} \quad T\,dS = dU + P\,dV \qquad \ldots(3.13)$$

This is the combined statement of the first and the second laws of thermodynamics. This combined statement includes only those terms which are state functions and hence is exact differential equation in terms of the internal energy and the entropy. This relation is therefore independent of path and depends on state properties only.

For irreversible processes second law gives

$$dS > 0$$

or therefore

$$T\,dS > dU + P\,dV \qquad \ldots(3.14)$$

On combining the Eqs. (3.13) and (3.14) gives

$$T\,dS \geq dU + P\,dV \qquad ...(3.15)$$

This is the most useful form and is most extensively used for further deductions. A reversible process is very very close to equilibrium and hence the equality sign in the above equation indicates equilibrium relationship. The inequality sign in one direction indicates spontaneous or natural process as shown above. By the same logic if

$$T\,dS > dU + P\,dV \qquad ...(3.16)$$

indicates unnatural process and energy will have to be invested to make it go.

3.5 VARIATION OF ENTROPY WITH TEMPERATURE (CONSTANT VOLUME)

This dependency of entropy on temperature can be evaluated under constant volume conditions as follows :

Second law gives $dS = \partial q/T$

putting value of ∂q from first law equation

$$dS = \frac{dU}{T} + \frac{P}{T}.dV$$

Putting value of dU in terms of partial derivatives as given in Eq. (2.20) and using

$$dU = C_P\,dT$$

leads to

$$dS = \frac{C_V}{T}dT + \left[P + \left(\frac{\partial U}{\partial V}\right)_T\right]\frac{dV}{T} \qquad ...(3.17)$$

at constant volume $dV = 0$

it reduces to

$$(dS)_V = \frac{C_V}{T}dT = C_V\,d\ln T$$

and a finite change

$$(\Delta S) = \int_{T_1}^{T_2} C_V\,d\ln T \qquad ...(3.18)$$

or

$$S_{T_2} = S_{T_1} + \int_{T_1}^{T_2} C_V\,d\ln T \qquad ...(3.19)$$

where S_{T_1} is entropy at some reference temperature T_1. As a corollary it can be deduced that :

By definition itself entropy has something to do with heat exchanged by the system. Hence it $\partial q = 0$ then the entropy change has to be zero.

At constant volume conditions when $dV = 0$ the first law equation

$$dU = \partial q - P\,dV \quad \text{reduces to} \quad dU = \partial q = T\,dS$$

or

$$\left(\frac{dU}{dS}\right)_V = T \qquad ...(3.20)$$

or if dq is zero, there is no entropy change in the system then

$$dU = -P\,dV$$

or

$$\left(\frac{dU}{dV}\right)_S = -P \qquad ...(3.21)$$

In other words there can be **conditions of constant entropy or sc to say adiabatic conditions when the internal energy change is equivalent to mechanical work exchange.**

3.6 VARIATION OF ENTROPY WITH TEMPERATURE (CONSTANT PRESSURE)

Now $(dS)_P = \partial q/T$ by definition

putting $\partial q = dU + P\,dV = d(U + PV) = dH$

$$(dS)_P = \left(\frac{dH}{T}\right)_P = \frac{1}{T}\left(\frac{dH}{dT}\right) dT = \frac{C_P}{T}\,dT$$

or for a finite change $(\Delta S)_P = \int_{T_1}^{T_2} C_P \, d \ln T$

or at some temperature T_2

$$S_{T_2} = S_{T_1} + \int_{T_1}^{T_2} C_P \, d \ln T$$

or

$$S_{T_2} = S_{T_1} + \int_{T_1}^{T_2} \frac{C_P}{T} . dT \qquad ...(3.22)$$

■ **Example 3.1** Calculate the entropy change when supercooled aluminium freezes at 800°K alongwith that in surroundings at the same temperature.

Solution : The process can be imagined to occur as shown in chapter 2 as Fig. 2.6. The solution of this requires the sp. heats of solid and liquid aluminium and H_f value for aluminium. The ΔH values are thus known and from this the entropy changes at various stages can be evaluated and then algebraically summated to get the entropy change in the system and the surroundings separately.

3.7 VARIATION OF ENTROPY WITH VOLUME (ISOTHERMAL)

The mathematical proof for evaluating the variation of entropy with volume under isothermal conditions is considered unnecessary at this stage and hence only the final form is given below :

If the thermal volume coefficient of expansion is denoted as

$$\alpha = \frac{1}{V}\left(\frac{\partial V}{\partial T}\right)_P \qquad ...(3.23)$$

and similarly the compressibility as

$$\beta = -\frac{1}{T}\left(\frac{\partial V}{\partial T}\right)_P \qquad ...(3.24)$$

then

$$dS = \frac{\alpha}{\beta}\,dV \qquad ...(3.25)$$

or

$$\Delta S = \frac{\alpha}{\beta}\int_{V_1}^{V_2} dV \qquad ...(3.26)$$

alternatively

$$S_{V_2} = S_{V_1} + \int_{V_1}^{V_2} dV \qquad ...(3.27)$$

the solution of this equation requires the knowledge of entropy value at some known volume as it will not be zero.

3.8 VARIATION OF ENTROPY WITH PRESSURE

It can also be shown that

$$dS = -\left(\frac{\partial V}{\partial T}\right)_P . dP \quad ...(3.28)$$

or

$$dS = -\alpha V dP \quad ...(3.29)$$

3.9 THERMODYNAMIC EQUATION OF STATE

It is possible to show that

$$P = T\left(\frac{\partial P}{\partial T}\right)_V - \left(\frac{\partial U}{\partial V}\right)_T \quad ...(3.30)$$

and also

$$V = T\left(\frac{\partial V}{\partial T}\right)_P + \left(\frac{\partial H}{\partial P}\right)_T \quad ...(3.31)$$

which correlate P, V and T and hence are referred to as additional thermodynamic equations of state.

The mathematical derivations are omitted here as these are considered unnecessary at the introductory level of the subject.

3.10 INTER-RELATION OF C_P AND C_V

It can be shown that

$$C_P - C_V = \frac{\alpha^2 VT}{\beta} \quad ...(3.32)$$

where α and β have the usual meaning as described in the previous section.

This relation is useful particularly because α and β are often known and it is rather easy to experimentally determine the values of C_P and hence knowing these C_V can be evaluated and likewise the experimental data of C_V and C_P may also be verified.

3.11 ENTROPY CHANGES WITH PHASE TRANSFORMATION

The evaluation of changes in entropy values associated with phase transformations requires the knowledge of heat exchanges associated with these processes like, latent heat of fusion of vaporisation or sublimation, crystal transformation and so on.

Melting : $\Delta S = \frac{q}{T} = \frac{H_f}{M.P.}$

Vaporisation : $\Delta S = \frac{q}{T} = \frac{H_V}{B.P.}$

Solid state changes : $\Delta S = \frac{H_S}{\text{temp. of transformation}}$

In a generalised may the entropy change in heating solid iron from say 500°C to 1600°C can be evaluated by a complex equation involving all possible changes from 500°C to 1600°C. Iron undergoes transformation as α to β at 1033°K, β to γ at 1183°K, γ to δ at 1673°K and melting at 1812°K. The overall equation giving the total entropy change in this process is as follows :

$$S_{1873°K} = \int_{773}^{1033} \frac{C_P}{T} dT + \frac{\Delta H(\alpha - \beta)}{1033} + \int_{1033}^{1183} \frac{C_P(\beta)}{T} dT + \frac{\Delta H(\beta - \gamma)}{1183}$$

$$+ \int_{1183}^{1673} \frac{C_P(\gamma)}{T} dT + \frac{\Delta H(\gamma - \delta)}{1673} + \int_{1673}^{1812} \frac{C_P(\delta)}{T} dT + \frac{\Delta H(liq)}{1812} + \frac{C_P(liq)}{T} dT \quad ...(3.33)$$

Or in a generalised from the total entropy of a substance at a certain higher temperature T can be evaluated from a equation of the type :

$$\Delta S = \Delta S_{298°} + \int_{298}^{T_1} \frac{C_P(\alpha)}{T} dT + \frac{H_f}{T_1} + \int_{T_1}^{T_2} \frac{C_P(\beta)}{T} dT + \text{and so on} \quad ...(3.34)$$

till the temperature under consideration is reached. This is graphically shown in Fig. 3.4. In a schematic manner and the solution of which needs the knowledge of entropy value at 298°K (25°C). These values are available as $\Delta S_{298°K}$ (like the earlier $\Delta S_{298°K}$ in compiled form).

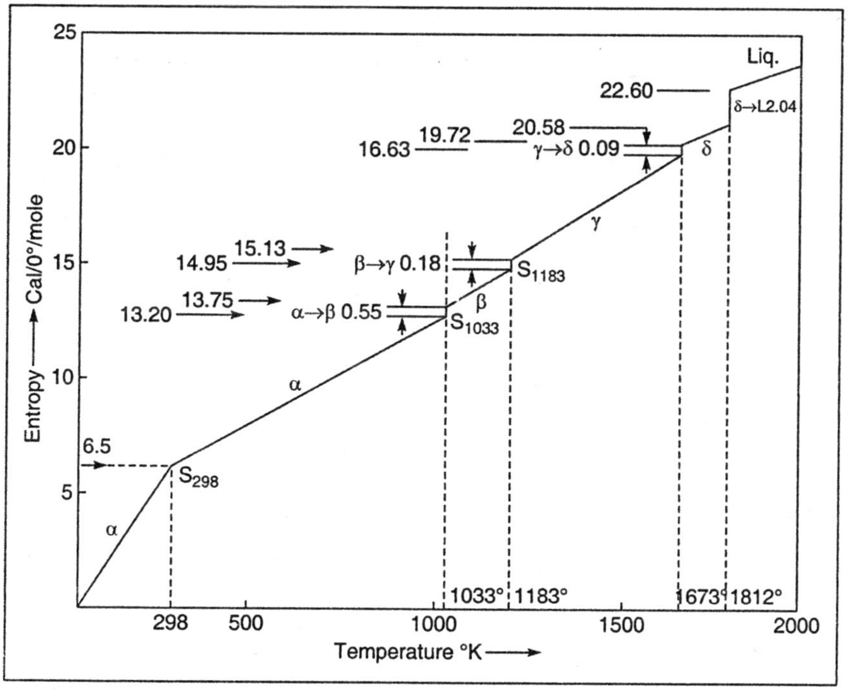

Fig. 3.4 Variation of entropy on heating iron from 25°C to above its melting point. The $S^{\circ}_{298°K}$ is taken as 6.5 cal/deg/mole. Note the absence of linear relationship between temperature and the entropy values although it has been assumed to be so in plotting this relationship.

■ **Example 3.2** Calculate the change in entropy when 2 kg of solid silver at 400°C is added to already molten 5 kg of silver at 1100°C. Find out the entropy change in the process.

Given : C_P (solid) = 5 cal/0°/mole

C_P (liq) = 7.3 cal/0°/mole

H_f (Ag) = 2690 cal/mole at 960°C (mp.)

Solution : The solution of this requires that first the final temperature of the total mass of seven kg of silver is determined. For this the process of mixing of the two has to be imagined as shown in Fig. 3.5. It means that first ($\Delta H_1 + \Delta H_2 + \Delta H_3$) shall be released to the surroundings, when the original 5 kg silver comes down to 400°C temperature. Then the mixing is carried out without any enthalpy change. The mass is now 7 kg. Whatever enthalpy was earlier given to the surrounding is now pumped back to the system. Since it is not certain as to what extent the heating will take place, the heating is carried out in stages. First the mass is heated to the melting point 960°C. It the available enthalpy is not enough even to heat upto this temperature, to whatever temperature it is capable of heating shall be the final temperature. But if the enthalpy is surplus then the remaining is used to melt the silver. Depending upon the surplus the equivalent silver will be melted while the other shall be still solid. In that case the final temperature shall be the melting point 960°C.

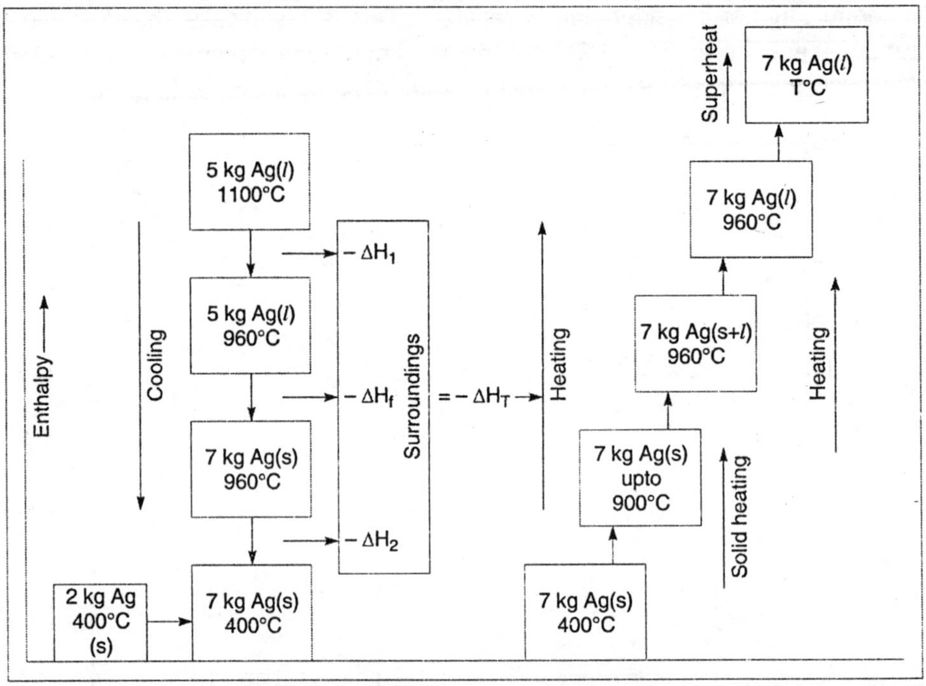

Fig. 3.5 Shows the imaginary route of carrying out the thermodynamic process to obtain desired result.

But even after melting all the mass of silver, enthalpy is still left then it will be used to superheat silver and the final temperature can be calculated. All these calculations require the specific heat values of solid and liquid silver and its latent heat of melting and which are given.

Once the final temperature is found out and if it is the melting point then the amount of solid left will be calculated. Then the entropy changes can be evaluated knowing the enthalpy changes in each stage. Their algebraic summation shall give the required value.

3.12 ENTROPY CHANGE ASSOCIATED WITH PROCESSES

For a reversible process the system receives heat from its surrounding and gains entropy and if it looses heat to the surroundings it looses entropy, amounting to the heat exchanged divided by the temperature. In this process it must be noted that the surrounding also takes part in the process in terms of either loss of gain of the same amount of entropy as the case may be for the first law has to be obeyed. The net sum of the entropy change of the system and the surrounding is therefore always zero as :

$$\Delta S \text{ (system)} + \Delta S \text{ (surrounding)} = 0 \quad \text{...(3.35)}$$

for a reversible process.

If the process is irreversible or natural then in the light of Eq. (3.10)

$$\Delta S \text{ (system)} + \Delta S \text{ (surrounding)} > 0 \quad \text{...(3.36)}$$

and conversely if the reversible process is carried out in a system isolated from the surroundings then

$$\Delta S \text{ (system)} > 0 \quad \text{...(3.37)}$$

Also if the process is at equilibrium then since there is no heat exchange no entropy change will occur and hence

$$\Delta S \text{ (system)} = 0 \quad \text{...(3.38)}$$

Now, if a system undergoes a change from A to B then the entropy change is given by

$$\Delta S = S_B - S_A \quad \text{...(3.39)}$$

assuming certain restrictions of conditions.

If $S_A = S_B$ then $\Delta S = 0$ or it means that the system A – B is at equilibrium and will not undergo a change.

If $S_B > S_A$ then $\Delta S > 0$ and the process occurs spontaneously or in a natural way from A to B.

Alternatively, if $S_A > S_B$ then $\Delta S < 0$ and the process can not occur unless helped by external energy source. All these above equalities or inequalities are for the total values of entropies, *i.e.,* system plus surroundings. Under certain conditions therefore the second law of thermodynamics provides in the form of entropy change, a criterion to judge the equilibrium or non-equilibrium position in a process.

3.13 ENTROPY CHANGES IN CHEMICAL PROCESSES

Like enthalpy change as predicted by the Hess's law, entropy being extensive energy state function, shall be subject to Hess's law. Therefore, entropy change in a given process is the same irrespective of the path followed in bringing about the change, *i.e.,* in one step or in multiple steps.

Entropy of any substance can be evaluated as from Eq. (3.22), as

$$\Delta S = \int_{T_1}^{T_2} \frac{C_P \, . dT}{T}$$

The change in entropy is often related to datum level of ΔS_{298° values which have been established for a large number of elements and compounds and hence are available in tabular forms.

Hence, $$\Delta S_T = \Delta S_{298^\circ} + \int_{298}^{T} \frac{C_P}{T} \, . dT \quad \text{...(3.40)}$$

Therefore, entropy change in a generalised reaction

$$A + B = C + D$$

shall be

$$\Delta S^0_{298°} = \Delta S(C)_{298°} + \Delta S(D)_{298°} - \Delta S(A)_{298°} - \Delta S(B)_{298°}$$

If the reaction is carried out at any other elevated temperature then, as was done while deriving the Kirchoff's law for evaluation of enthalpy, the process can be imagined to take place at 298°K and the products raised to reach reaction temperature T or, reactants raised to temperature T and then reaction carried out. Therefore, to the above if the entropy of heating is taken into account the entropy of reaction can be calculated.

Therefore, $$\Delta S_{T_2} = \Delta S_{T_1} + \int_{T_1}^{T_2} \frac{\Delta C_P}{T} . dT \quad ...(3.41)$$

where ΔC_P has the same meaning as given in Kirchoff's law.

■ **Example 3.3** Calculate the entropy change for the reaction

$$Fe_2O_3 + 3C = 2Fe + 3CO$$

at 298°K.

Solution : Given : The standard values of entropies at 298°K for various compounds are :

FeO	21.4 cal/O/mole
C	49.5 cal/O/mole
Fe	32.63 cal/O/mole
CO	47.3 cal/O/mole

At 298°K $$\Delta S = 2S_{Fe} + 3S_{CO} - 3S_C - S_{Fe_2O_3}$$

$$= 65.26 + 141.9 - 148.5 - 21.4$$

$$= 37.26 \text{ cal/O°/mole}$$

3.14 CONCLUDING REMARKS

It is not possible to convert all heat into work without bringing some permanent change in the surroundings. Because of exchange of heat the system undergoes change in entropy - a measure of randomness of the system. Natural processes occur with increase in entropy. The second law provides method of calculating only difference in entropy of a system. The entropy change is an indication of direction of the change as to whether it is natural, unnatural or is at equilibrium. **Is this the only parameter that indicates the direction of a change?** The answer is given in the next chapter.

□□□

Chapter 4

Thermodynamic Potentials

Free Energy and Equilibrium Criteria

4.1 INTRODUCTION

A system by itself, in isolated state, or in contact with surroundings shall stay in equilibrium unless acted upon it by some constraints. But if a system tends to move, as a result of being not in equilibrium or as a result of external constraints, there must be a driving force making the system move from within itself or under the applied constraints. This driving force is referred to as thermodynamic potential driving the system to change to a new state. Is this driving force or potential the same under all conditions of the system or it varies from condition to condition of the system? In this context as heat flows from higher to a lower temperature the flow is possible due to the higher thermal potential in the form of higher temperature. Similarly, the higher pressure is a driving mechanical potential. Also higher electrical voltage is driving the electrons under the influence of electrical potential as voltage. These are more tangible examples and can be understood more readily. However, the question remains as to what is that potential, forcing a thermodynamic process, wherein energy exchange occurs, to take place? It is easy to imagine that a system with higher associated energy will be relatively unstable as compared to the one wherein the associated energy is lower. In other words when a natural change occurs the system moves from a higher to a lower energy level or moves from a higher to a lower thermodynamic potential. It also means that energy is a potential driving the system for change to occur. It also means that for equilibrium to exists the potentials of all the systems or sub-systems within it must be at the same potential.

So far the internal energy, enthalpy and entropy have been evolved as energy parameters but it is not yet known as to whether these act singly or in combination as potentials in driving a particular process?

In deriving various relationships from second law of thermodynamics entropy was seen to suggest that for a natural process it is more than zero, *i.e.,* +ve and therefore for a unnatural process it is less than zero, *i.e.,* –ve. Therefore, for equilibrium to exist the entropy shall have to be the same for all the systems and sub-systems existing together. But it must be noted here that for a natural change the system moves from a lower to a higher entropy level and hence entropy can not be considered as driving force bringing about the change. Can internal energy and enthalpy quality as potentials capable of driving the process towards equilibrium, if already at higher levels? If the internal energy and the enthalpy are potential terms then changes in them will have to be zero at equilibrium and these will have to decrease for natural changes or irreversible processes and increase for unnatural processes. Let us therefore, see if criteria can be evolved mathematically to evaluate what constitutes as thermodynamic potential or potentials and if so under what conditions?

4.1.1 Potential Under Constant Volume and Constant Entropy Conditions

These are peculiar conditions in as far as constant entropy is concerned, since constant volume processes are well known. The constancy of entropy, *i.e.*, isentropic condition is that where $dS = 0$, *i.e.*, $\partial q/T = 0$ and since T is generally not zero (except the extreme case of absolute zero state) therefore ∂q will have to be zero. In other words the process is adiabatic and no exchange of heat between the system and its surrounding is possible under the situation under consideration. The combined statement of the first and the second laws of thermodynamics as stated earlier in Eq. (3.13) that

$$dU = T\,dS - P\,dV$$

and putting $dS = 0$ and $dV = 0$ it reduces to

$$dU = 0$$

where $dU = 0$ for reversible or equilibrium process and $dU < 0$ for natural or irreversible process and $dU > 0$ for an unnatural process. This is based on the earlier observation that $dS > 0$ for natural and $dS < 0$ for unnatural processes.

The equilibrium for such a process under constant volume and constant entropy conditions shall be defined as

$$dV = 0, \quad dS = 0, \quad q = 0 \quad \text{and} \quad dU = 0$$

and for a natural process the internal energy will decrease, *i.e.*, will change from a higher to a lower level and for unnatural process just the reverse.

The internal energy thus fully qualifies to be called a potential under constant volume and entropy conditions. It also means that the equilibrium shall be characterised, under the constant volume and entropy conditions, highest possible entropy. As a corollary therefore the internal energy must be a function of entropy and volume and can be expressed as

$$U = F(S, V) \qquad \text{...(4.1)}$$

which on partial differentiation shall give

$$dU = \left(\frac{\partial U}{\partial S}\right)_V dS + \left(\frac{\partial U}{\partial V}\right)_S dV \qquad \text{...(4.2)}$$

The internal energy is also given by the combined statement of the first and second laws as

$$dU = T\,dS - P\,dV$$

on comparison of Eq. (4.2) with this leads to

$$\left(\frac{\partial U}{\partial S}\right)_V = T \quad \text{and} \quad \left(\frac{\partial U}{\partial V}\right)_S = -P \qquad \text{...(4.3)}$$

In other words by specifying U as a function of S and V it is possible to evaluate T and P and thereby describing the system fully. By expressing U as a function of any other two variables it will not be possible to describe all the variables, as is possible, under the specification of U, as a function of S and V. For example by stating say

$$U = F(P, V)$$

it will not be possible to determine the remaining properties S and T of the system. The reader should now understand why such peculiar conditions were assumed in the beginning in deriving the internal energy as a potential term.

Now that since internal energy as one of the forms of energies of a system can be related to all the variables of the system, any change in the properties of the system can be directly correlated to the energy exchange and that is what is required.

From the combined statement of the first and second laws as $dU = T\,dS - P\,dV$ it can be readily inferred that in a reversible isochoric process change in internal energy is given by

$$\int dU = T \int dS = dq \quad \text{...(4.4)}$$

or by the reversible heat exchange. Also in a reversible adiabatic process, *i.e.*, when $\partial q = T\,dS = 0$ or under isentropic conditions the change in internal energy is equal and opposite of reversible work done on the system as

$$\int dU = -P \int dV \quad \text{...(4.5)}$$

4.1.2 Potential Under Constant Pressure and Constant Entropy Conditions

Now, therefore, it should be clear that these conditions are so chosen such that the complete system is describable.

The conditions here are that $dP = 0$ and $dS = 0$.

Now, $$H = U + PV$$

therefore on differentiation

$$dH = dU + P\,dV + V\,dP$$

by putting in first law and second law equations

$$dH = q + V\,dP = T\,dS + V\,dP \quad \text{...(4.6)}$$

At equilibrium when $dP = 0$ and $dS = 0$

$$dH = 0$$

In a natural process since $T\,dS > 0$

$dH > 0$ for a natural process

and since $T\,dS < 0$ for an unnatural process

$dH > 0$ for an unnatural process

So, therefore on the whole

$$dH > = < 0$$

for unnatural, equilibrium and natural processes as the case may be. Therefore, the term **enthalpy qualifies as being called a potential term under constant pressure and entropy conditions.** The enthalpy can therefore be described mathematically as

$$H = F(P, S)$$

On partial differentiation the enthalpy change can be described as

$$DH = \left(\frac{\partial H}{\partial S}\right)_P . dS + \left(\frac{\partial H}{\partial P}\right)_S . dP \quad ...(4.7)$$

Comparing this with already derived equation (4.6)

$$\left(\frac{\partial H}{\partial S}\right)_P = T \quad \text{and} \quad \left(\frac{\partial H}{\partial P}\right)_S = V \quad ...(4.8)$$

Again by expressing H as a function of P and S for any system the remaining variables T and V can be readily found out and thus the system gets completely described.

It is also clear from the equation (4.6) that for isobaric processes the change in enthalpy of a process is nothing but reversible heat exchange since $dH = T\,dS = \partial q$ and for adiabatic processes when $q = T\,dS = 0$ the change in enthalpy will $\int V\,dP$ over the pressure limits of initial and final states.

4.1.3 Potential Under Constant Volume and Constant Temperature Conditions

These are useful conditions under which many practical processes are carried out in the form of a fixed enclosure and a fixed temperature. These conditions mathematically are

$$dV = 0 \quad \text{and} \quad dT = 0$$

and as $T\,dS = dU + P\,dV$ for a reversible isochoric process

therefore, $(dU - T\,dS) = 0$

or $d(U - TS) = 0$

Let us define a new mathematical term as

$$A = U - TS$$

then $dA = 0$ for a reversible process

$dA < 0$ for a natural process

and $dA > 0$ for a unnatural process.

This newly coined function **A as defined by (U – TS) qualifies for being referred to as a potential term under constant volume and constant temperature conditions**. Being a function of variables which are all state functions 'A' must also be a state function.

Expressing as before $A = F(T, V)$

and by partial differentiation

$$dA = \left(\frac{\partial A}{\partial T}\right)_V . dT + \left(\frac{\partial A}{\partial V}\right)_T dV \quad ...(4.9)$$

Now, $A = U - TS$

therefore, $dA = dU - T\,dS - S\,dT$

and since $T\,dS = dU + P\,dV$

therefore, $dA = -S\,dT - P\,dV \quad ...(4.10)$

and comparing the two leads to

$$\left(\frac{\partial A}{\partial T}\right)_V = -S \quad \text{and} \quad \left(\frac{\partial A}{\partial T}\right)_T = -P \quad ...(4.11)$$

As A varies with T and V and if this is known then S and P of the system can be evaluated to describe the system fully.

This function A was first defined by Helmholtz and therefore is known as **Helmholtz's Free Energy.** From the definition of A it can be written after differentiation as

$$dA = dU - T\,dS \qquad \text{(at constant temperature)}$$

and for a reversible process $q_{rev} = T\,dS$

and therefore, $q_{rev} = dU - dA$

According to the first law, $dU = q_{rev} - w_{rev}$

and hence $q_{rev} = dU + w_{rev}$

comparing the two equations leads to

$$-dA = w_{rev}$$

or $dA = -w_{rev}$

Change in A during a process is given by

$$\Delta A = -w_{rev}$$

That the change in Helmholtz free energy in a process is equal to the amount of work done by the system on to the surroundings or is equal to the amount of work the system is capable of doing (being reversible it is the maximum work possible). The function A is therefore also known as **'work function'**. If the system is capable of doing mechanical work alone then the loss in mechanical work is equal to the change in Helmholtz's free energy. The loss in A is equal to the maximum work the system is capable of doing as a result of given isothermal change and that being maximum, will have to be achieved through reversible change only.

From $dA = -P\,dV - S\,dT$ as in Eq. (4.10)

it is clear that for isothermal processes

$$\int dA = -P \,.\int dV \qquad \text{...(4.12)}$$

or the change in free energy is given by maximum work the system is capable of doing on the surroundings by way of volume expansion. Also for isochoric processes

$$\int dA = -S \,.\int dT$$

and free energy change will be in terms of temperature variation of the system.

4.1.4 Potential Under Constant Pressure and Constant Temperature Conditions

These are by far the most commonly adopted conditions in chemical and metallurgical engineering practices.

It means $dP = 0$ and $dT = 0$

The combined statement of the first and second laws of thermodynamics can be written as

$$d\,(TS) = dU + d\,(PV) \qquad \text{...(4.13)}$$

or $d\,(U + PV - TS) = 0$

Now let us define a new term mathematically as

$$G = (U + PV - TS) \quad ...(4.14)$$

and therefore obviously dG = 0 for reversible or process at equilibrium

and since $dS > 0$ for natural process

$U + PV - TS < 0$ for natural process and hence

$dG < 0$ for natural process.

Conversely and obviously

$dG > 0$ for an unnatural process

The property or function **G** as defined above thus **qualifies for being called as potential term under constant pressure and constant temperature conditions**. The function G is a variable of functions which are all state functions and hence G also is a state function. This function G is referred to as **'Gibb's Free Energy'** as distinct from the Helmholtz free energy.

Now, $G = U + PV - TS$

or putting value of $H = U + PV$

$$\Delta G = H - TS \quad ...(4.15)$$

or for a finite change $\Delta G = \Delta H - T\,\Delta S$

or $dG = dH - T\,dS - S\,dT$

and at constant pressure $dP = 0$

$$dH = q = T\,dS$$

and hence $dG = -S\,dT$

or

$$\left(\frac{dG}{dT}\right)_P = -S \quad ...(4.16)$$

Again $dG = dH - T\,dS - S\,dT$

on putting $dH = dU + P\,dV$

$$dG = dU + P\,dV + V\,dP - T\,dS - S\,dT$$

putting in this $dU + P\,dV = T\,dS$

$$dG = T\,dS + V\,dP - T\,dS - S\,dT$$

or

$$= V\,dP - S\,dT \quad ...(4.17)$$

at constant temperature when dT = 0

$$dG = V\,dP$$

or

$$\left(\frac{dG}{dP}\right)_T = V \quad ...(4.18)$$

Since G is a state function any change in G can be represented as

$$\Delta G = \Delta U - T\,\Delta S + \Delta(PV)$$

$$= -w_{rev} + \Delta(PV)$$

or for small changes

$$dG = -w_{rev} + d(PV) \quad ...(4.19)$$

Therefore, the change in Gibb's free energy is equal to negative reversible work obtainable from the thermodynamic process (as is Helmholtz free energy) but corrected for change in volume during the process, *i.e.,* corrected for mechanical work done during the process. In the definition itself it is so as would be obvious as

$$A = U - TS$$

and $$G = U - TS + PV \quad \text{(the PV term is the correction factor)}$$

By adopting G as a function of T and P as

$$G = F(P, T) \qquad \ldots(4.20)$$

on partial differentiation

$$dG = \left(\frac{\partial G}{\partial T}\right)_P dT + \left(\frac{\partial G}{\partial P}\right)_T dP$$

and since $$dG = -S\,dT + V\,dP \qquad \text{as in Eq. (4.17)}$$

on comparison

$$\left(\frac{\partial G}{\partial T}\right)_P = -S \quad \text{and} \quad \left(\frac{\partial G}{\partial P}\right)_T = V \qquad \ldots(4.21)$$

Since all are exact differentials ∂G in place of dG makes no change and the same applies for others.

Thus by knowing G as a function of T and P it is possible to establish other two variables *viz.* S and V and obtain the description of the system fully.

■ **Example 4.1** Calculate the free energy change of a reaction

$$<Si> + \{O_2\} = <SiO_2>$$

at a temperature 900°K.

Solution : $\Delta G^0_{900°} = \Delta H^0_{900°} - T\,\Delta S^0_{900°}$

$$= \left[\Delta H^0_{298} + \int_{298}^{900} \Delta C_P\, dT\right] - T\left[\Delta S^0_{298} + \int_{298}^{900} \frac{\Delta C_P}{T}.dT\right]$$

where $\Delta C = \Sigma\, C_{P\,(products)} - \Sigma\, C_{P(reactants)}$

Therefore, to solve this the following information is required either as directly given or indirectly calculable.

(1) C_P values of individual reactants and products in the range of 298°–900°K.

(2) $\Delta H°$ and $\Delta S°$ values for all reactants and products at 298°K as standard values.

If these are available then the above equation can be solved and the ΣG°_{900} can be obtained.

4.1.5 Functions U, H, A and G and their Extensive Nature

All these are mathematically defined functions and are interrelated to one another and to P, V, T and S as

$$H = U - PV$$

$$A = U - TS$$

$$G = U - TS + PV = A + PV$$
$$= U + PV - TS = H - TS$$

All these are useful to evaluate energy exchanges in thermodynamic processes. Since energy of a system can be related to the variables of the system, any change in the properties of the system can directly be correlated to the energy exchange and that is what is required. It should also be noted that out of all these energy parameters only G is such that it can be assessed in terms of all variables such as P, V, T and S at a time. The function A can also be similarly correlated as $A = q - PV - TS$, but it is not exact differential equation. The functions U and H are relatable to only some of these four variables. G is therefore by far the most exhaustive energy term that can account for varied variations in a system. No wonder it is used extensively to assess the energy exchanges in chemical and metallurgical processes wherein not only P, V and T change but also the chemistry. All these four properties are extensive state properties and are additive. For any process involving chemical and/or physical changes the change in the free energy is given by the difference in the free energies of the two state, one before and the other after the process has occurred. Therefore, $\Delta G = \Delta H - T\,\Delta S$ and since H and S have been shown to be measured with reference to a reference state, so G also has to be assessed in reference to such a reference state. Hence like H and S at the reference level

$$\Delta G° = \Delta H° - T\,\Delta S°$$

Similarly, the Helmholtz free energy A also can be expressed with a reference level.

4.1.6 Availability of Work/Energy from Systems

How much energy and in whatever form, is extractable from a system? It cannot be any amount but it is fixed once and for all from a well defined system in a well defined surroundings. Let us say that water is stored at a level of M_1 meters from the mean sea level and the hydro-electric generator is located at M_2 meters from m.s.l. The energy that is available in the form of potential energy to be converted to electric power is equivalent of $(M_1 - M_2)$ meters. From the same dam if generators are located at M_3 which is less than M_2 or at M_4 level which is more than M_2 level the available potential from the same reservoir is either increased (at M_3) or decreased (if at M_4). This is also obvious from the definition of energy that it is not absolute but equivalent of effects it produces on the surroundings. It is measured as a difference and upto the level to which it can be lowered. It can not be lowered below that of the surroundings.

Now let us imagine a system at pressure P_s and temperature T_s in contact with surroundings at P and T. Let us also assume that the system exchanges heat and work with the surroundings. Let us assume that q amount of heat is given by the surroundings to the system and

therefore, $q = \text{ or } < (T_s\,\Delta S)$

Now, since $\Delta U = q - w = \text{ or } < (T_s\,\Delta S - P_s\,\Delta V)$

where work done is only mechanical and on the surroundings only. By transposing the terms

$$\Delta U - T_s\,\Delta S + P_s\,\Delta V = \text{ or } < 0 \quad ...(4.22)$$

The Gibb's free energy has been defined as above on the left hand side and hence from the above

$$\Delta G = \text{ or } < 0$$

This is the energy freely available in the form of heat and mechanical work from the system or its **'availability'** (disregarding energy changes in all other forms). It may be surprising to see that the energy available is a function of T_S and P_S and not P and T, *i.e.*, the functions of the surrounding conditions and not of the system itself.

The inequality sign indicates non-reversible condition and the equality sign the reversible condition when maximum work will be carried out. Hence taking the equality for the change

$$\Delta G = \Delta U - T_S\ \Delta S + P_S\ \Delta V \quad \text{...(4.23)}$$

putting $\Delta U = q - w = T\ \Delta S - P\ \Delta V$ in the above

$$\Delta G = T\ \Delta S - T_S\ \Delta S - PDV + P_S\ \Delta V$$

$$= (T - T_S)\ \Delta S - (P - P_S)\ \Delta V \quad \text{...(4.24)}$$

The term $- T\ \Delta S$ is the heat extracted from the system and can be gainfully converted to work through a Carnot engine operating between T and T_S temperatures. The work done by the Carnot engine would then be $-(T - T_S)\ \Delta S$. Similarly, if the system is allowed to expand as ΔV the mechanical work done by the system shall be effectively $(P - P_S)\ \Delta V$. Therefore, the above equation (4.24) shows the free energy or the maximum work (energy in generalised term) available from the system.

If $P = P_S$ and $T = T_S$ then no network is available as $\Delta G = 0$. The system is then in equilibrium with the surrounding and hence no exchange of energy is possible.

The larger are $(P - P_S)$ and $(T - T_S)$ gaps the larger will be the energy available from the system. In chemical reactions when chemical bonds change, the energy exchange is taken care of by the term $- T\ \Delta S$.

It is thus clear that maximum energy available from a system is not absolute but it is with reference to the conditions of surroundings. In a given surroundings ΔG gives the maximum exchange possible and hence very aptly named as **free energy** of the system.

In the above example of hydroelectric generation the level of the dam is the surrounding, source of energy, and the level of the hydro-electric generator as the system.

If the temperature and the pressure of the surroundings are fixed the availability of a given system is fixed.

4.2 PARTIAL PROPERTIES AND CHEMICAL POTENTIAL

It is easy to appreciate extensive properties like volume, internal energy, enthalpy, entropy free energy, etc. for pure substances in terms of unit mass or per mass mole (as molal property). But when the substance is in dissolved form, *i.e.*, in an homogeneous multicomponent system, what is its molal extensive property? The answer is not simple. This can be understood by taking an example of volume for the purpose of illustration. This is chosen because it can almost be **'seen'** and thus readily understood. When in solution the individual component occupies the total volume of the solution as its own volume irrespective of its concentration in solution. Can it then be said that the molal volume of the component in solution is equal to the solution volume? If this is assumed then the molal volume is practically the same for all concentrations of all the components in solution. This is absurd.

Similarly, it is rather difficult to visualise the potential term driving a chemical reaction between phases wherein more than one components are involved. The question as to which way a component would move between the two phases under a given set of conditions needs to be answered in the same way as answered above while deriving conditions for determining the potential term. Under the characteristic conditions of constancy of S and V, internal energy acts as the potential driving the process. Under these constancy conditions does U become the potential term driving the chemical component from one to the other phase? Same argument can be made for H, A and G values also. The answer is not direct and is interlinked with the molal internal energy of that component in the multicomponent system. The molal property of a component in a multicomponent system thus needs to be properly understood.

The problem is solved indirectly in terms of thermodynamic concepts by coining a new term called **'partial quantity or property'**. This is applicable to extensive properties only.

Let us imagine the components, solute and solvent, initially in separate vessels as shown in Fig. 4.1(a). Each is one g. mole in mass. Let us imagine that the solute is transferred to the solvent but somehow kept separate as shown in Fig. 4.1(b). In the next step the partition is removed and the

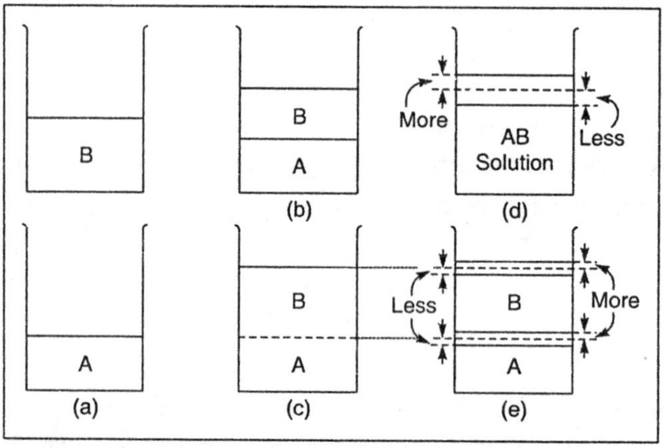

Fig. 4.1 The process of mixing of two components to form a solution with some imaginary steps.

two are allowed to get dissolved to form a homogeneous system - the solution. It is almost rare that the solution volume will be equal to the algebraic sum of the component volumes. In reality the solution volume is either less or more than the algebraic sum of their molal component volumes as shown in Fig. 4.1(c). Now let us imagine that whatever be the individual sum total of the atomic or molar volumes of the components in solution, these are somehow isolated in the form of two layers, the algebraic sum being equal to that of the system volume. Whatever is the individual component molal volume, the molal volume in the system is either less or more than that value when in pure form. This change is due to the interaction of the components when present together. The interaction originates from their different chemical nature, electronic structure and so on.

Although the above picture can not be constructed in reality it can be evaluated by mathematical formulations as follows :

While deriving the thermodynamic potential terms it has already been shown that

$$U = F(S, V); \quad H = F(S, P); \quad A = F(T, V) \quad \text{and} \quad G = F(T, P)$$

These are valid for systems of fixed composition and when characteristic two of the four variables P, V, T and S are fixed.

For example at constant T and P the system attains equilibrium at a minimum value of G. But if composition is changed keeping T and P the same, being an extensive property the value of G gets changed and hence the system no longer remains in equilibrium. The same arguments can be made for U, H and A values as per above equations. Hence strictly speaking if chemistry of the system is variable the above equations get modified as :

$$U' = F(S, V, n_1, n_2, n_3, ..., n_i) \quad ...(4.25)$$

$$H' = F(S, P, n_1, n_2, n_3, ..., n_i) \quad ...(4.26)$$

$$A' = F(T, V, n_1, n_2, n_3, ..., n_i) \quad ...(4.27)$$

$$G' = F(T, P, n_1, n_2, n_3, ..., n_i) \quad ...(4.28)$$

The prime on these functions is to indicate the overall property of the solution as distinct from that related to only one mole or atom. The following is illustrated for G only but the same treatment is valid for other properties like U, H and A.

Differentiation of Eq. (4.28) gives :

$$dG' = \left(\frac{\partial G}{\partial T}\right)_{P, n_1, n_2, n_3, ..., n_i} . dT + \left(\frac{\partial G}{\partial P}\right)_{T, n_1, n_2, n_3, ..., n_i} . dP + \left(\frac{\partial G}{\partial n_1}\right)_{T, P, n_2, n_3, ..., n_i} . dn_1 + \left(\frac{\partial G}{\partial n_2}\right)_{T, P, n_1, n_3, ..., n_i} . dn_2 \ldots + \left(\frac{\partial G}{\partial n_i}\right)_{T, P, n_1, ..., n_{i-1}} dn_i \quad ...(4.29)$$

If composition is fixed, except the first two terms on the right hand side all other vanish. This equation can be simplified further as

$$dG' = -S\,dT + V\,dP + \sum_{n_1}^{n_i} \left(\frac{\partial G}{\partial n_i}\right)_{T, P, n_1, n_2, ..., n_{i-1}} . dn_i \quad ...(4.30)$$

In this the last terms mean the rate of change of values of G' with changes in concentration of one component at a time. How is it possible to change concentration of one component in a system and yet not change that of the others in the system?

This difficulty is overcome by imagining a system large enough such that by addition of one mole of a component i would not materially alter the composition of the other components.

The change $\partial G'$ accompanying this one mole addition is thus the molal G value of that component i in the multicomponent system. It is expressed in the differential form as

$$\left(\frac{\partial G}{\partial n_i}\right)_{T, P, n_1, n_2, ..., n_{i-1}} \quad ...(4.31)$$

as the molal property of component i in the system consisting of several other components. This is known as **'partial property of that component'**. This is often denoted by putting a bar on the symbol as for example $\overline{G}$ called as G-bar or bar-G. Therefore,

Partial molal internal energy $\overline{U} = \left(\frac{\partial U}{\partial n_i}\right)_{S, V, n_1, n_2, \ldots, n_{i-1}}$ ...(4.32)

Partial molal enthalpy $\overline{H} = \left(\frac{\partial G}{\partial n_i}\right)_{S, P, n_1, n_2, n_3, \ldots, n_{i-1}}$ ...(4.33)

Partial molal Helmholtz free energy $\overline{A} = \left(\frac{\partial A}{\partial n_i}\right)_{T, T, n_1, n_2, \ldots, n_{i-1}}$ ...(4.34)

and the partial molal Gibb's free energy is as shown in Eq. (4.31).

The total extensive property of a homogeneous system, comprised of more than one component, consists of partial parts contributed by each one of the component, thereby making the total property of the system. This individual contribution of a component to the overall total extensive property is called its **'partial property'** and when expressed for a total mass of one gram mole or atom in the system becomes its **'partial molal property'**, and when expressed for a fraction of the total one gram mole or atom of solution is its **'partial molar property'**. The partial property is ordinarily different from its pure phase property. The variation depends upon the interaction between dissimilar components in the system.

At constant T and P the free energy of a solution would be :

$$dG' = \sum_1^i \left(\frac{\partial G}{\partial n_i}\right)_{n_1, n_2, n_3, \ldots, n_{i-1}} = \sum_1^i \partial \overline{G}_i \quad \ldots(4.35)$$

For a given composition with minimum G' the system shall be in equilibrium. If the concentration of one component is changed, obviously depending upon the factual values either G' will decrease or increase as given by the equation above. The system shall gain equilibrium by allowing exchange of that component with the surroundings or with another sub-system till again the original equilibrium composition is achieved. In other words the value of $\overline{G}_i$, *i.e.*, partial molar free energy is acting as a potential in driving that chemical component in a certain direction to achieve equilibrium. The partial molar free energy $\overline{G}_i$ is thus the potential in terms of chemical composition of the system. This is therefore designated as **'chemical potential'** and expressed commonly by the letter μ in the literature. By arguing similarly for U, H and A it can be stated as :

$$\mu = \left(\frac{\partial U}{\partial n_i}\right)_{S, V, n_1, n_2, n_3, \ldots, n_{i-1}} = \left(\frac{\partial H}{\partial n_i}\right)_{S, P, n_1, n_2, n_3, \ldots, n_{i-1}}$$

$$= \left(\frac{\partial A}{\partial n_i}\right)_{T, V, n_1, n_2, n_3, \ldots, n_{i-1}} = \left(\frac{\partial G}{\partial n_i}\right)_{T, P, n_1, n_2, n_3, \ldots, n_{i-1}} \quad \ldots(4.36)$$

or $\mu = (\overline{U}_i)$ or $(\overline{H}_i)$ or $(\overline{A}_i)$ or $(\overline{G}_i)$ ...(4.37)

However, **each term shall act as chemical potential term under the conditions of constancy under which it acts as the thermodynamic potential term.** It must therefore be understood that U is a characteristic function of independent variables S and V and composition and likewise others as discussed earlier. These can be written as :

$$dU = T\,dS - P\,dV + \sum_{1}^{i} \mu_i \,.\, dn_i \qquad ...(4.38)$$

$$dH = T\,dS + V\,dP + \sum_{1}^{i} \mu_i \,.\, dn_i \qquad ...(4.39)$$

$$dA = -S\,dT - P\,dV + \sum_{1}^{i} \mu_i \,.\, dn_i \qquad ...(4.40)$$

$$dG = -S\,dT + V\,dP + \sum_{1}^{i} \mu_i \,.\, dn_i \qquad ...(4.41)$$

Now, since from first law $dU = q - w$

from Eq. (2.12) $= T\,dS - P\,dV - w'$

therefore, $dU = T\,dS - P\,dV - $ [chemical work in the process (w')] ...(4.42)

Comparison of Eqs. (4.38) and (4.42) give that chemical work is $= \sum_{1}^{i} \mu_i \,.\, dn_i$, and it means that migration of chemical species over the available coordinate when multiplied by the driving force that is the chemical potential, μ_i in this case gives the chemical work done in the process by the migration of the chemical species.

Partial molar energies as shown above are the driving chemical potentials. If a two, phase multicomponent system exists, then the $\mu_{i\,(phase\ i)} = \mu_{i\,(phase\ ii)}$ for equilibrium to exist for component i. In case these are not similar then i will migrate from higher to a lower potential side.

Like all other potential terms the chemical potentials of a component in different phases in equilibrium or between system and its surroundings or between sub-system of a system will have to be equal for the component to stay in equilibrium. If it is unequal in that case the higher chemical potential shall drive the component from higher to a lower potential zone.

Likewise complete chemical equilibrium shall be possible only if chemical potentials for all components are correspondingly the same in all zones of the system. Transfer of chemical species from a zone of a higher potential to a lower potential shall be a natural process and *vice-versa* as a unnatural process.

Most of the chemical and metallurgical processes are carried out under constant temperature and pressure and therefore partial molar free energy $\overline{G}_i$ is commonly used as chemical potential term to assess chemical equilibria.

4.2.1 Maxwell's Relations

We have derived

$dU = T\,dS - P\,dV$ as in Eq. (3.13)

$dH = T\,dS + V\,dP$ as in Eq. (4.6)

$dA = -S\,dT - P\,dV$ as in Eq. (4.10)

$dG = -S\,dT + V\,dP$ as in Eq. (4.17)

All these are exact differential equations.

From the property of exact differentials the following relationships can be stated :

$$\left(\frac{\partial T}{\partial V}\right)_S = -\left(\frac{\partial P}{\partial S}\right)_V \quad ...(4.43)$$

$$\left(\frac{\partial T}{\partial P}\right)_S = -\left(\frac{\partial V}{\partial S}\right)_P \quad ...(4.44)$$

$$\left(\frac{\partial S}{\partial V}\right)_T = -\left(\frac{\partial P}{\partial T}\right)_V \quad ...(4.45)$$

$$\left(\frac{\partial S}{\partial P}\right)_T = -\left(\frac{\partial V}{\partial T}\right)_P \quad ...(4.46)$$

These are popularly known as **'Maxwell's relations'** and are useful for calculations of thermodynamic parameters. There are four variables *viz.* P, V, T and S and the variation of one with the other is related to the variation of the third with the fourth variable under certain conditions, in the relationships. Thermodynamic **Mnemonic square** is available to arrive at these relationships by a thumb rule. These equations apply to any system which is determined by two parameters and the remaining two are determinable, even otherwise as seen before.

Also from the above stated relationships for dU, dH, dA and dG it can be derived directly for fixed chemical composition the following relationships :

$$T = \left(\frac{\partial U}{\partial S}\right)_V = \left(\frac{\partial H}{\partial S}\right)_P \quad ...(4.47)$$

$$P = -\left(\frac{\partial U}{\partial V}\right)_S = -\left(\frac{\partial A}{\partial V}\right)_T \quad ...(4.48)$$

$$V = \left(\frac{\partial H}{\partial P}\right)_S = \left(\frac{\partial G}{\partial P}\right)_T \quad ...(4.49)$$

and

$$S = -\left(\frac{\partial A}{\partial T}\right)_V = -\left(\frac{\partial G}{\partial T}\right)_P \quad ...(4.50)$$

4.3 GIBB'S HELMHOLTZ EQUATION

It is known that $G = H - TS$

putting values of S as in Eq. (4.16)

$$G = H + T\left(\frac{\partial G}{\partial T}\right)_P$$

dividing by T^2

$$\frac{G}{T^2} - \frac{1}{T}\left(\frac{\partial G}{\partial T}\right)_P = \frac{H}{T^2}$$

i.e.,

$$\frac{1}{T}\left[\frac{G}{T} - \left(\frac{\partial G}{\partial T}\right)_P\right] = \frac{H}{T^2}$$

or

$$\left[\frac{\partial (G/T)}{\partial T}\right]_P = -\frac{H}{T^2} \quad \text{or} \quad \frac{\partial (G/T)}{\partial (1/T)} = H$$

or $$\left[\frac{\partial(\Delta G/T)}{\partial(1/T)}\right]_P = \Delta H \qquad ...(4.51)$$

Similarly, it can be derived that

$$\left[\frac{\partial(\Delta A/T)}{\partial(1/T)}\right] = \Delta U \qquad ...(4.52)$$

The Eqs. (4.51) and (4.52) are known as **Gibb's Helmholtz Equations** for free energies as they vary with temperatures.

4.4 FUGACITY

It has been shown earlier that for a system doing only mechanical work

$$dG = V\,dP - S\,dT$$

and at constant temperature when $dT = 0$

$$dG = V\,dP$$

For one mole of an ideal gas

$$V = RT/P$$

therefore, $$dG = RT\,.\,dP/P = RT.\,d\ln P \qquad ...(4.53)$$

It can be imagined that a gas mixture can be in equilibrium with only one component across a diaphram which is permeable to that component only. Hence in this system

$$dG\text{ (pure component side)} = d\overline{G}_i\text{ (mixture side)}$$

or therefore, $dG = RT\,d\ln P$ where P is the partial pressure of that component on the mixture side.

This can be integrated over known limits of G and the corresponding pressures as

$$G_2 - G_1 = RT\ln(P_2/P_1) \qquad ...(4.54)$$

and if conditions for G are to be taken as agreed standard in which P_1 will be one atmosphere then

$$G_2 = G° + RT\ln P_2 \qquad ...(4.55)$$

For an ideal mixture of gases this will take the form as

$$\overline{G}_i = G_i° + RT\ln p_i \qquad ...(4.56)$$

it can also be imagined that an ideal gas mixture can be in partial equilibrium with only one component across a diaphragm which is permeable to that component only. Hence in this system

$$dG_{(\text{pure component side})} = dG_{(\text{mixture side})} \qquad ...(4.57)$$

or therefore, $$d\overline{G}_i = RT\ln p_i$$

where p_i is the total pressure on pure side of p on the mixture side. On integration it again leads to

$$\overline{G}_i = G_i° + RT\ln p_i \qquad ...(4.58)$$

Strictly speaking this equation is applicable to ideal gases or their mixtures only. The free energy-pressure relationship is linear in that case. But for the real gases as pressure rises the integration of this equation does not remain linear and poses some difficulties. **Lewis**, the famous thermodynamics scientist introduced a new term called **'fugacity'** and mathematically defined it as

$$d\overline{G}_i = RT\,d\ln f_i \qquad ...(4.59)$$

On integration $RT\,d\ln f = G + I$

where I is the integration constant and it is so chosen that

$$f = p \quad \text{as} \quad p \to 0$$

Mathematically the functions U, A and H can also be correlated in a similar way as :

Since $dH = T\,dS + V\,dP$ as in Eq. (4.6)

or at constant entropy $= V\,dP$

and since $V = RT/P = RT\,d\ln P = RT\,d\ln f$ ...(4.60)

As $dA = -S\,dT - P\,dV$ as in Eq. (4.10)

at constant temperature $= -P\,dV$

$$= -RT\,(dV/V) = RT\,d\ln V$$

therefore, $A = \int_{V_i}^{V_f} d\ln V$ ...(4.61)

As $dU = T\,dS - V\,dP$ as in Eq. (3.13)

$= -V\,dP$ under constant entropy

$$= -RT\,(dP/P) = -RT\,d\ln P$$

$= -RT\,d\ln f$ ...(4.62)

The introduction of such a function ensures linear relationship between all these extensive properties, U, H, A and G and logarithm of fugacity in any state of any gas.

For an ideal gas, fugacity is equal to pressure at all pressures. Similarly, for gas mixture the fugacity of each component equals partial pressure, as the total or also partial pressure tends to be zero. The fugacity of a component in an ideal gas mixture is numerically equal to the partial pressure of that component.

For an ideal or real gas component the pressure is mean of the fugacity and the pressure it would have if it had behaved as an ideal gas.

Therefore, $P \to 0$ then $f = P$

and $P \to 0$ then $f/P = 1$

If a gas mixture is separated by a diaphragm from one of the constituent gas on the other side and depending upon the fugacity values of that component on either side of the diaphragm the diffusion of that component across the diaphragm will occur if

$RT\,d\ln P\,(\text{pure}) \neq RT\,d\ln P_i$ ...(4.63)

or $RT\,d\ln f\,(\text{pure}) \neq RT\,d\ln f_i$ ...(4.64)

and in the direction from higher to lower fugacity side as is obvious like mechanical equilibrium in terms of pressures.

Therefore, fugacity represents the escaping tendency of the component in a system. Fugacity values can be calculated from above relationships using equation of state and compressibility data and its variation with temperature and pressure.

Fugacity is also a potential term like any other term defined earlier indicating natural process, unnatural process and equilibrium state.

Solids and liquids do have vapour pressures which are very low and hence can be equated to as the fugacities in gas phase in equilibrium with solid/liquid as the case may be and hence by equilibrium considerations solids and liquids do have fugacities.

4.5 ACTIVITY

Activity of a component is defined as its ability to interact with the surroundings. The activities of pure substances is therefore maximum. This ability is maximum when fugacity is maximum. The activity has therefore been defined as the ratio of fugacity of the substance in the state in which it exists to the fugacity in its standard state as

$$a = \frac{f_i}{f^\circ} \quad \text{or} \quad a_i = \frac{f_i}{f_i{}^\circ} \qquad \text{...(4.65)}$$

where **f°** and $\mathbf{f_i}$ are the fugacities of pure component and the ith component. The standard state is commonly pure liquid or solid at one atmosphere pressure and at the temperature under consideration or gas at one atmospheric pressure at the temperature under consideration. Obviously only one, solid or liquid or gas phase, is stable at the given temperature, excepting the transition temperature. The fugacity in the standard state is normally described by the symbol **f°** and hence activity is unity for pure substances. Since standard state is arbitrary it is not necessary to take pure substances alone as standard states. It is common practice to take 1 wt% dil solution as standard state dealing with metallic solutions.

With the above

$$a = (f/f^\circ) \quad \text{or} \quad a_i = (f_i/G_i{}^\circ)$$

and hence

$$f = f^\circ . a \quad \text{or} \quad f_i = a_i\, G_i{}^\circ$$

and since f° and $f_i{}^\circ$ are fixed and constant, therefore,

$$RT\, d\ln f = RT\, d\ln a = dG \qquad \text{...(4.66)}$$

and

$$RT\, d\ln f_i = RT\, d\ln a_i = d\overline{G}_i \qquad \text{...(4.67)}$$

The similarity in activity and fugacity is obvious.

Integrating from the standard state at $a = 1$ at $f = f^\circ$

or

$f_i = G_i{}^\circ$ to any arbitrary state

$$RT \ln a = G - G^\circ \qquad \text{...(4.68)}$$

and

$$RT \ln a_i = \overline{G}_i - G_i{}^\circ \qquad \text{...(4.69)}$$

and on differentiating

$$RT\, d\ln a = d\,(G - G^\circ) \qquad \text{...(4.70)}$$

and

$$RT\, d\ln a_i = d(G_i - G_i{}^\circ) \qquad \text{...(4.71)}$$

This activity is also a potential term since it is almost proportional to fugacity and the latter being potential term so activity also should be a potential term. It indicates that if activities on either side of an interface of two phases in a system are the same, then the system shall be in equilibrium. If the two are not same on either side of the interface then the component with higher activity shall have a tendency to migrate to the region of lower activity zone across the boundary, if physically otherwise possible.

4.6 CONCLUSIONS

The parameters like internal energy, enthalpy, work function and Gibb's free energy have been the potential terms under their characteristic conditions. These energy properties are extensive properties. Potentials are intensive properties and the differentiation of these two must be properly understood. The properties like temperature, pressure and chemical potential are intensive properties and are potential terms of one kind or the other and which make the heat flow (under thermal potential gradient), make work exchange possible (under pressure gradient) and make chemical species move (under chemical potential gradient). The values of the potential terms decide whether the system is at equilibrium or is poised for a spontaneous change or whether energy will have to be supplied to make the system move in the desired direction.

The terms fugacity and activity are directly related to each other. If say two immiscible liquids like a slag and a metal or like oil and water are in equilibrium with respect to certain component the activity of that component in both the phases must be the same. In other words the fugacities also must be the same and since and which in turn means that their chemical potentials in the two phases must also be the same. Therefore the terms fugacity, activity, partial energy properties under the characteristic conditions and finally the chemical potential means the same potential term, only in different forms, that dictate chemical equilibria in thermodynamic systems.

□□□

Chapter 5

Third Law of Thermodynamics

Estimation of Entropy

5.1 ZERO ENERGY LEVEL

The first law allows measurement of changes in internal energy and enthalpy of a system when thermodynamic processes occur, though not in absolute terms but only relatively and hence the zero levels for internal energy or enthalpy are purely arbitrary. Both of these are energy parameters. The entropy defined by the second law is also a energy parameter. The second law allows measurement of only entropy changes in a thermodynamic process but the entropy value in absolute terms is not determinable from the second law.

The Carnot cycle leads to a relationship

$$T_2/T_1 = q_2/q_1$$

which states that the difference in temperatures T_1 and T_2 are fixed when q_1 and q_2 amounts of heats are then exchanged with the surroundings respectively. The range of temperature scale over T_1 and T_2 is independent of the nature of the system in the Carnot cycle. It can as well be fixed as on 0° to 100° Celcius scale by adjusting the real values of q_1 and q_2. The same scale could then be extended to lower and higher ranges of the same Celcius scale. From this it comes out to be –273°C as zero degrees absolute or 0°K. As both energy and entropy decrease with falling temperature will energy and entropy reduce to zero at absolute zero temperature?

All pure substances can not be assigned a zero energy value at absolute zero because then the heat of formation of compounds will have to be zero at absolute zero and which is just not the case. This is also obvious from the fact that though internal energy changes with temperature it can not be extrapolated to zero at absolute zero temperature. Energies at higher temperatures can be estimated experimentally from $\int C_P\, dT$ or $\int C_V\, dT$ and these can be extrapolated to absolute zero temperature. But these have not been observed to be zero at 0°K. Thus any assignment of zero value of internal energy to even elements at absolute zero temperature is purely arbitrary.

The integration of Gibbs-Helmholtz equations

$$d\,(\Delta A/T) = -\,\Delta U/T^2\, dT$$

and

$$d\,(\Delta G/T) = -\,\Delta H/T^2\, dT$$

gives separately
$$\frac{\Delta A}{T} = -\int_{T_1}^{T_2} \frac{\Delta U}{T^2} dT + I_V \quad ...(5.1)$$

and
$$\frac{\Delta G}{T} = -\int_{T_1}^{T_2} \frac{\Delta H}{T^2} dT + I_P \quad ...(5.2)$$

where I_V and I_P are integration constants indicating the constant volume or the constant pressure conditions of the systems. These constants I_V and I_P can not be determined from first or the second laws of thermodynamics and empirical methods alone become necessary for their evaluations.

5.2 ZERO ENTROPY LEVEL

The entropy variation with temperature has already been shown earlier in terms of the specific heats as

$$dS_V = C_V \frac{dT}{T}$$

and
$$dS_P = C_P \frac{dT}{T}$$

and on integration leads to

$$S_{T_V} = S_{O_V} + \int_0^T C_V \frac{dT}{T} \quad ...(5.3)$$

$$S_{T_P} = S_{O_P} + \int_0^T C_P \frac{dT}{T} \quad ...(5.4)$$

where S_{O_V} and S_{O_P} are integration constants or values of entropy at absolute zero temperature under constant volume and pressure conditions respectively. Again second law by itself can not solve this for evaluation of S_{O_V} and S_{O_P} values except by direct experimental work.

It was Nernst who after analysing the data of variation of ΔG and ΔH with temperature, particularly at lower temperature region, found that $\Delta G/T$ and $\Delta H/T$ tends to zero as the temperature approaches zero value as schematically shown in Fig. 5.1.

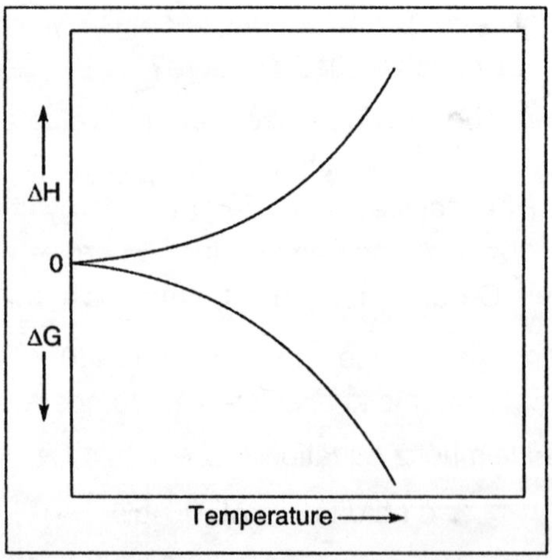

Fig. 5.1 Showing schematically the observations of Nerns't about variation of ΔH and ΔG of reactions with temperatures.

$$\frac{\partial(\Delta G)}{\partial T} = 0 \text{ and } \frac{\partial(\Delta H)}{\partial T} = 0$$

$$T \to 0 \qquad T \to 0$$

since
$$\frac{\partial(\Delta G)}{\partial T} = -\Delta S, \quad \Delta S = 0 \qquad \text{...(5.5)}$$

$$T \to 0$$

Nernst generalised these findings in the form of a statement that **for all reactions in which substances are in condensed state ΔS is zero at the absolute zero temperature.** This is known as **Nernst's Heat Theorem** and is one of the early statements of the third law of thermodynamics. Hence if reaction occurs as

$$A + B = AB$$

the entropy change of the reaction is given by

$$\Delta S_{(reaction)} = \Delta S_{AB} - \Delta S_A - \Delta S_B$$

and if the entropies of elements are assigned zero values as per the above theorem then the entropy of the compound AB is also zero at the absolute zero temperature. This is in fact what the Nernst Heat Theorem states.

The combined statement of the first and the second laws of thermodynamics is

$$dG = V\,dP - S\,dT$$

and this can take the form

$$d\,\Delta G = \Delta V\,dP - \Delta S\,dT$$

or
$$\frac{\partial(\Delta G)}{dT} = \Delta V \frac{dP}{dT} - \Delta S$$

at constant pressure
$$\left(\frac{\partial(\Delta G)}{dT}\right)_P = -\Delta S = 0 \qquad \text{...(5.6)}$$

$$T \to 0$$

Similarly,
$$\Delta G = \Delta H - T\,\Delta S$$

differentiating partially with respect to temperature at constant pressure

$$\left(\frac{\partial(\Delta G)}{\partial T}\right)_P = \left(\frac{\partial(\Delta G)}{\partial T}\right)_P - T\left(\frac{\partial \Delta S}{\partial T}\right) - \Delta S \qquad \text{...(5.7)}$$

putting values of

$$\left(\frac{\partial(\Delta G)}{\partial T}\right)_P = -\Delta S \text{ and } \left(\frac{\partial(\Delta H)}{\partial T}\right)_P = \Delta C_P$$

therefore,
$$\Delta C_P = T\left(\frac{\partial \Delta S}{\partial T}\right)_P \qquad \text{...(5.8)}$$

Hence, $\Delta C_P = 0$ at $T = 0$ unless $T\left(\frac{\partial \Delta S}{\partial T}\right)_P$ is infinite and which is not at all so.

According to Debye's law

$$C_P \text{ or } C_V \propto T^3 \quad \text{or} \quad C_P \text{ or } C_V \propto kT^3 \text{ where k is constant}$$

hence at $T = 0$
$$C_P \text{ or } C_V = 0$$

Hence the specific heats are zero at absolute zero temperature and hence the isothermal processes like change in pressure and/or volume, since without any heat exchanges as C_P and C_V are zero, therefore, occur without variation in entropies.

or $$\left(\frac{\partial \Delta S}{\partial P}\right)_T = 0 \qquad ...(5.9)$$

and $$\left(\frac{\partial \Delta S}{\partial V}\right)_V = 0 \text{ as } T \to 0 \qquad ...(5.10)$$

or in other words the entropy attains a value of zero at absolute zero temperature.

5.3 THIRD LAW OF THERMODYNAMICS

The third law may now be stated as that the **'entropy of any homogeneous substance which is in complete internal equilibrium may be taken as zero at zero degree absolute temperature'**. Complete internal equilibrium means that unique lowest energy state which is obtained in the crystals when cooled under perfect equilibrium conditions, such that the atoms occupy uniquely ordered arrangement and hence the uniquely lowest energy state.

There is no theoretical proof for the first and the second laws of thermodynamics but for the third law proof can be given in terms of quantum statistics.

It should be noted that, similar to internal energy and the enthalpy, the entropy of elements at the uniquely lowest energy level is zero at absolute zero temperature, is again arbitrary. This is however fruitful because this also, as a consequence, leads to that entropy of a compound at absolute zero temperature is also zero, provided that the compound is in complete internal equilibrium.

It is rather difficult to achieve the state of complete internal equilibrium and hence atoms in compounds are not that orderly arranged and hence may not have zero but some entropy which might cause error in calculations.

5.4 EVALUATION OF ENTROPY AT HIGHER TEMPERATURES

The variation of entropy of a substance with temperature has been shown in Section 3.5 and 3.6 as

at constant volume $$S_{T_2} = S_{T_1} + \int_{T_1}^{T_2} . C_V . d \ln T \qquad ...(5.11)$$

at constant pressure $$S_{T_2} = S_{T_1} + \int_{T_1}^{T_2} . C_P . d \ln T \qquad ...(5.12)$$

If this integration is carried from 0° to a temperature T°K then using third law these equations would reduce to :

at constant volume $$S_{T_2} = \int_0^T C_V \, d \ln T \qquad ...(5.13)$$

at constant pressure $$S_{T_2} = \int_0^T C_P \, d \ln T \qquad ...(5.14)$$

If this is split into two different temperature zones one lower than the Debye's characteristic temperature θ_D (which is generally around 15°K) and the other above this temperature then,

$$\Delta S_{T_2} = \int_0^{\theta_D} \frac{C_P}{T} dT + \int_{\theta_D}^{T_2} \frac{C_P}{T} dT \quad ...(5.15)$$

putting value of $C_P = kT^3$ as per Debye's T^3 law

then
$$\Delta S_{T_2} = \int_0^{\theta_D} \frac{kT^3}{T} dT + \int_{\theta_D}^{T_2} \frac{C_P}{T} dT$$

$$= \int_0^{\theta_D} \frac{kT^2}{T} dT + \int_{\theta_D}^{T_2} \frac{C_P}{T} dT$$

$$= \frac{1}{3} . kT^3_{\theta_D} + \int_{\theta_D}^{T_2} \frac{C_P}{T} dT$$

as the first term vanishes at T = 0° K

$$\Delta S_{T_2} = \frac{1}{3} C_{P(\theta_D)} + \int_{\theta_D}^{T_2} \frac{C_P}{T} dT \quad ...(5.16)$$

or where the first term on the right hand side is the heat capacity at Debye's characteristic temperature. This can be solved graphically by plotting either C_P/T vs T or C_P vs ln T.

The area between the points θ_D and T_2 under the curve bound by the temperature axis shall be the value of the second term on the right hand side and when $1/3.C_{P(\theta_D)}$ is added to it shall give the entropy change required. Th... presumes that upto temperature T_2 no phase change occurs. If phase change occurs in between, before T_2 then the equation will have to be extended as in Eq. (3.41).

In a generalised form the values of $S_{298°K}$ are known from the values of C_P between the temperature range 0°–298°K and are available in tabular form for many substances and hence entropy of a substance at a temperature T may be evaluated as :

$$S^0_T = S^0_{298°} + \int_{298°}^{T} C_V . \frac{dT}{T} = S^0_{298°} + \int_{298°}^{T} C_V \, d \ln T \quad ...(5.17)$$

$$S^0_T = S^0_{298°} + \int_{298°}^{T} C_P . \frac{dT}{T} = S^0_{298°} + \int_{298°}^{T} C_P \, d \ln T \quad ...(5.18)$$

the superscript indicates standard states. If there is a phase change in between then the entropy change will be given by as discussed in Section(3.11).

■ **Example 5.1** Calculate the standard molar entropy of zinc at 300°C assuming that it follows Debye's equation below 25°K. Assume average mean specific heat for solid zinc as 0.36 cal/g/0°K and that at 25°K is 0.091 cal/0°/gram.

Solution :
$$\Delta S_{300°C} = \Delta S^0_{573°K} = \int_{0°}^{25°} \Delta S + \int_{25°}^{573°} C_P \, . d \ln T$$

$$= \frac{1}{3} C_{P(25°)} + C_{P(ave)} \times [\ln 573 - \ln 25]$$

$$= 0.03 + 1.931 = 1.961 \text{ cal/0°/gram}$$

$$= 1.961 \times 65.5 = 132 \text{ cal/0°/g.mole}$$

It is generally not so accurate to take mean value of specific heat like this over a temperature range. In fact C_P value in this temperature range must be evaluated by experiment and then the integral 25–573°K is evaluated by plotting C_P vs. ln T and the area between the limits 25° and 573°K shall be taken as the value of that integral. This is more precise.

5.5 CONSEQUENCES OF THIRD LAW

The coefficient of thermal expansion and compression tends to be zero as temperature approaches absolute zero. This will be obvious from Maxwell's relations. That is

$$\alpha = \frac{1}{V}\left(\frac{\partial V}{\partial T}\right)_P$$

substituting Eq. (4.46) in this gives

$$\alpha = -\left(\frac{\partial S}{\partial P}\right)_T \frac{1}{P} = 0 \quad \text{as S tends to zero as } T \to 0 \qquad \text{...(5.19)}$$

$$\alpha = \frac{1}{P}\left(\frac{\partial P}{\partial T}\right)_V$$

substituting Eq. (4.45) in this gives

$$\beta = \left(\frac{\partial S}{\partial V}\right)_T \frac{1}{P} = 0 \quad \text{as S tends to zero as } T \to 0 \qquad \text{...(5.20)}$$

It is also true that the third law itself suggests impossibility of attaining absolute zero temperature. This can be proved thus :

Let us imagine a Carnot cycle which is operating between 0° and T°K. The entropy change during such a cycle is zero and hence the entropy change of the cycle :

$$\frac{q_1}{T_1} + \frac{q_0}{T_0} = 0$$

The term q°/T° at zero absolute is zero and hence q_1/T_1 must also be zero. But this can not be so because it is a finite quantity and hence the only alternative conclusion possible is that 0°K is not attainable. Therefore, it has been stated that **'no definite series of processes could lead to achieve absolute zero temperature'**. The third law therefore finds a way out to estimate absolute values of entropy based on quantum statistics.

The assignment of a zero value to the entropy of uniquely stable arrangement of elements is nothing but convention. The assignment of zero value to the entropy of uniquely stable compounds is a consequence of the above convention and is stated as the third law of thermodynamics.

5.6 CONCLUDING REMARKS

Third law finds a way out to estimate entropy of substances at absolute zero degree and thereby completes the method of calculating the entropy of substances. This ultimately is required to work out energy exchanges during thermodynamic processes.

□□□

Chapter 6

Zeroth Law of Thermodynamics

Phase Stability

6.1 INTRODUCTION

If a system is fully isolated by an adiabatic wall and left to itself, it attains a stable state such that no further change is perceptible, no matter how long one waits, then the system is said to be in equilibrium or has attained equilibrium. Left to itself it shall continue to be in the state of equilibrium and shall change from the equilibrium state only if externally acted upon by some constraint. A system may also be in equilibrium with its surroundings in the same way. If a system consists of only one phase, say liquid or solid or gas then left to itself, being a homogeneous phase it is normally in the state of equilibrium except some exceptional circumstances. This is is what is known as **'stable equilibrium'**. This is best illustrated by an example of a ball resting in a valley. A system in stable equilibrium shall get displaced from its equilibrium state on application of an external constraint, but as soon as the external constraint is removed the system shall gain its original stable state.

If such displacement, under external constraint makes it stable in some other state then it is said to be in **'metastable equilibrium'**. The best example to illustrate this meta-stability is the ball resting in a small depression at the top of a hill. The other better examples of metastability are the case of a supercooled liquid or a mixture of hydrogen and oxygen. A human body is most stable in horizontal lying position. Standing vertical, though appears quite stable, yet it is only meta-stability. In all these cases the system shall move to a stabler form if given a chance by initiating the displacement.

On the contrary, on displacement, if the system changes to a displaced state, it is said to be in **'neutral equilibrium'**. The example for its illustration is a ball resting on a flat plane surface.

In the chemical and metallurgical arena quite often it is necessary to know whether a certain phase is stable or not and if it is unstable, what is the stable phase under the given constraints. Similarly it is more often required to deal with two or more phases containing several components in contact with each other. It is also then necessary to know whether these phases are in stable equilibrium or not and if not, then what is the stable configuration? The former part is dealt with by ascertaining the conditions for phase or phases stability and the later part by knowing generally the chemical equilibria between the phases involved. The cases like two-phase-equilibria in a single component system like liquid-vapour or liquid-solid can be dealt with in terms of phase stability of an individual phase or phases involved in the system.

The thermodynamic studies are confined to stable equilibria only. It has already been pointed out that the system and its surroundings can be in a state of mechanical equilibrium, if pressures in both are equal or in thermal equilibrium if both are at the same temperatures. Similarly, they can be in a sate chemical equilibrium if the chemical potentials are the same in both. The same holds true within the system itself vis-a-vis the different phases involved or vis-a-vis the different sub-systems within the system itself under consideration.

The concept of mechanical stability is a static one as will be obvious from the example of overhead tank. On opening the tap at the lower level, water flows down and stays at that lower level, *i.e.*, it has acquired a static stability at the lower level.

In chemistry however the concept of stability is a dynamic one. In terms of the theory of reaction rates, the system acquires stability only when the rates of forward and that of the backward reaction exactly match to have no net effect whatsoever. The two equal and opposite rates match but do not reduce the system to no activity level. This is why it is a dynamic concept.

6.2 ZEROTH LAW OF THERMODYNAMICS

This law is more basic than even the first law, although it was discovered only after the first and the second laws were formulated and, formalised only after the enunciation of the third law of thermodynamics. In stead of giving it the number fourth it has been preferred to call it zeroth law thereby indicating its more elementary nature than even the first law. This law can be appreciated even without appreciating the idea of temperature derived from the second law.

Two basic properties, particularly of fluids and in that also, more easily appreciated in the case of gaseous phase, are 'pressure' and 'volume'. A cylinder and a piston can be imagined to contain a certain amount of gas. The volume and pressure of this gas can be adjusted by the movements of the piston. Left the values of pressure and volume for a certain gas in this cylinder be P_1 and V_1. The gas is thus defined in terms of P_1 and V_1. If another gas at volume V_2 is brought in contact with this gas through a diathermal wall the two can be brought to equilibrium by adjusting the pressure of the second gas P_2. The two gas systems may have the same hotness or coldness or fixed so by placing them in a certain oven or a refrigerator. Out of the four variables three may be taken arbitrarily but the fourth will have to be fixed, in relation to these three for equilibrium to reach. It means that if gas one at P_1V_1 is placed in a oven and gas two of V_2 is brought in contact in the same oven with diathermal walls between the two, the P_2 will have to be adjusted to bring the two gas systems in equilibrium.

Let us imagine a third gas which is brought in equilibrium with the first one in place of the second with P_3 and V_3 values such that P_1V_1 and P_3V_3 are now in equilibrium in the oven or the refrigerator. Now, from the pair **'one-three'** the first system is removed and the second is brought in, then the pair **'two-three'** must also be in equilibrium, as is obvious and observed in practice. The same need not be true with only gases but shall be true with any other bodies.

This observation is stated in the form of **'zeroth law of thermodynamics'** as :

'If of three bodies say A, B and C, B and C are separately in equilibrium with A then, B and C are also in equilibrium with one another'.

The converse is also true in that if three bodies are in contact, each to each and are all in equilibrium together, then any two taken separately are also in equilibrium with one another.

This leads to the fact that the function of P and V for a certain amount of gas is the empirical parameter hotness or coldness of the system and is denoted by the parameter T in the form of an equation as

$$F(P, V) = T$$

This then would be the equation of state of that fluid. It is this function of the fluids, which is nothing but temperature, which has the property of taking the same value for fluids in equilibrium with one-another. The fluid is thus uniquely defined by any two of the three variables P, V and T.

The original zeroth law statement therefore reads that of three bodies A, B and C, if A-B and A-C are in thermal equilibrium meaning thereby that at the same temperature, then B-C are also in thermal equilibrium, meaning thereby that, they are at the same temperature. Conversely this is possible for fluids only if their temperatures are equal, that is the product PV for each one of three fluids A, B and C are equal.

Similarly, if V_1T_1 of A are in equilibrium separately with V_2T_1 of B and V_3T_1 of C then V_2T_1 of B is in equilibrium with V_3T_1 of C as they mean same pressure for A, B and C; which shall be the case of mechanical equilibrium.

Equally true will be the case of chemical potentials. If chemical potentials of A and B are equal and that of A and C are also equal then they are in equilibrium and hence B and C also shall be in equilibrium from the chemical point of view. Since thermodynamic equilibrium means equilibrium with respect to all possible potentials, thermal, mechanical, chemical, etc., the generalised statement as above has been made as the zeroth law statement for an overall equilibrium.

It may also be stated that if three of more systems are in contact, each with each, and are in equilibrium with respect to any one or more of the thermodynamic potentials, then any two taken together shall also be in equilibrium with respect to that potential.

6.3 PHASE STABILITY

A substance can exist in either a solid, liquid or a gas form depending upon the conditions of pressure, temperature and volume. The knowledge of conditions under which a phase is stable, *i.e.*, in equilibrium is important in systematic thermodynamic analysis. This will also give rise to the conditions under which two or more phases exists together in equilibrium like at melting and boiling points. From the point of view of consideration of thermodynamic potentials, of all the possibilities, that phase which has the lowest associated potential shall be the stable one relative to the others. If the phase has a higher potential, it shall change over to the one having lower potential.

Single component phase is stable in its existing form if and only if the rate of change of entropy with its own mass is negative, *i.e.*,

$$\left(\frac{\partial s}{\partial n}\right) < 0$$

This is obvious because if this is positive then this will be a natural change possible and hence phase will be unstable.

Other conditions are that the specific heat values C_P and C_V must be positive and the compressibility β must always be positive, *i.e.*,

$$\beta > 0 \text{ or } -\frac{1}{V}\left(\frac{\partial V}{\partial P}\right)_T > 0$$

Free energy must vary with chemical concentration in a certain direction. It means activity must vary positively with concentration. If this variation changes or reverses the phase is unstable.

The problem of phase stability can best be understood from the example of liquifaction of a gas by compression coupled with lowering of the temperature. The Fig. 6.1 shows the P-V plots for a gas at various temperatures. It is seen that as the temperature is lowered the P-V isotherm is lowered in position until a temperature T_C – **called critical temperature** is reached, below which the gas liquifies. It means the gas stability is lost and liquid is formed for a certain range of P and V values at a temperature lower than the T_C. The gas is stable in the range A-C and B-F but not in between in the range C-F.

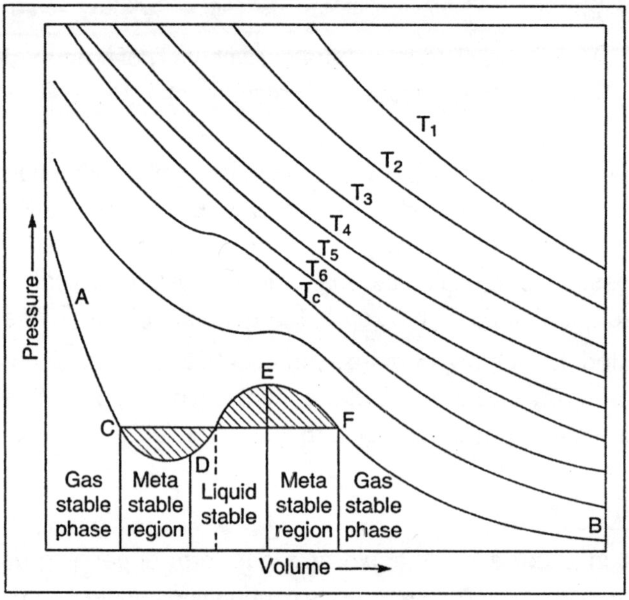

Fig. 6.1 Pressure-volume relationship for air as affected by temperature to show the conditions for liquification of air.

As the gas is cooled $(\partial V/\partial P)_T$ has a +ve value of compressibility upto temperature T_C at which a slight inflextion occurs in the P-V isotherm. Below T_C at D and E the slope is zero and the gas becomes unstable. A small range before this stage is the metastable area C-D and E-F. In the range D-E the compressibility is negative and hence unstable for gas phase to exist and hence the change to a more stabler phase that can exist, *i.e.,* liquification takes place.

Phase stability will be lost and phase transition shall take place if the free energy variation shows a discontinuity with temperature. In other words a stable phase region shows continuous variation of free energy with temperature. The first derivative of free energy as derived from the combined statement of the first and the second law as dG = V dP – S dT and hence

$$\left(\frac{\partial G}{\partial T}\right)_P = -S, \quad \left(\frac{\partial G}{\partial P}\right)_T = P \quad \text{and} \quad \left[\left(\frac{\partial (G/T)}{\partial (1/T)}\right)\right] = H$$

that is S, V and H vary continuously for the phase stability. Any **discontinuity is an indication of phase instability**. Similarly, the second derivatives $(\partial^2 G/\partial T^2)$ and the third $(\partial^2 G/\partial T^3)$ also be varying continuously with temperature for phase stability. Phase stability will be lost if heat input increases the temperature in a two phase stable system like solid-liquid or liquid-gas.

In magnetic transformation S, V and H vs. T are continuous curves but the second derivatives of free energy vs temperature is a discontinuous one.

Order-disorder transformation shall be possible if S, V and H show discontinuity in their variation with temperature. These changes do cause abrupt change in sp. heat, coefficient of thermal expansion and compressibility. These are second order manifestations of free energy variation with temperature.

6.4 TWO PHASE STABILITY/EQUILIBRIA

It is a common knowledge that at the melting point both solid and liquid and at the boiling point both the liquid and the gas co-exist. This is because the two phases solid and liquid in one case and the liquid and the gas in the other case are in thermal equilibrium. This is because the two phases in contact have the same thermodynamic potential at such critical temperatures. The change in free energy when one component gets transferred from one to the other phase is zero at such critical temperatures.

Since $\Delta G = \Delta H - T\,\Delta S$, at equilibrium $\Delta H = -T\,\Delta S$. The latent heat is enthalpy at constant pressure and temperature and the internal energy at constant temperature and volume. The heat input is latent head and matches $-T\,\Delta S$ and hence $\Delta G = 0$. The integral $\int P\,dV$ should be zero for such changes. In Fig. 6.1 the two thatched areas are equal and opposite and hence the total effective $\int P\,dV$ is zero when both liquid and gas coexist together. It can be mathematically proved that for two phases to be in equilibrium like liquid and solid, in any system, the temperature, the pressure and the free energy of both the phases will have to be in equilibrium and this will be possible if any only if the chemical potentials of all the components, *i.e.,* partial molal free energy values of all the components,though different values for different components, are equal for each component in both the phases.

6.4.1 Phase Rule

A phase is a homogeneous part of the system with definite boundaries and interfaces with other phases or sub-systems. The number of components in a system is the smallest number of substances in terms of which the composition of each phase in a system can be defined. The degrees of freedom are nothing but number of independent intensive variables to be controlled externally for describing the system completely like pressure, volume and temperature. The phase rule states that for any system

$$F = C - P + 2 \qquad \ldots(6.1)$$

or

$$F + P = C + 2 \qquad \ldots(6.2)$$

where F is the degrees of freedom, C the number of components involved and P the number of phases present. For systems normally encountered in the metallurgical area the pressure is often equal to atmospheric pressure and hence is fixed at one atmosphere and the systems are condensed phases and therefore, the phase rule takes the form as

$$F = C - P + 1 \quad ...(6.3)$$

because one degree of freedom in the form of pressure is not available.

6.4.2 P-G-T Relationships vs Phase Stability

On the whole in a closed one-component system at a certain temperature T and pressure P, the equilibrium phase or phases in the system shall be that or those having minimum value of free energy. The equilibrium thus can be evaluated by knowing the variation of G at constant T and P. But if any of these two also vary then through the relationship of G-P at constant T or G-T at constant P or as G-P-T relationships, as all the three vary, may be useful. Phase stabilities can also be appreciated from P-T diagrams but G-P, G-T and G-P-T diagrams are more appropriate.

The P-V relationship for a single component system as shown in Fig. 6.2. It is clear that at one atmosphere pressure upto melting point T_M solid is stable, above boiling point T_B gaseous phase is stable and in between T_M and T_B liquid is stable. At T_M both solid and liquid and at T_B both liquid and gas phases co-exist. As pressure is lowered the boiling point decreases rapidly and the melting point also decreases but rather slowly. Consequently at a certain temperature T_S all solid, liquid and gas phases co-exist. The pressure at which this happens is much below one atmosphere. At temperatures below T_S therefore solid and gas can co-exist together and hence sublimation is possible but above T_S solid will have to change to liquid and then to gas on increasing the temperature.

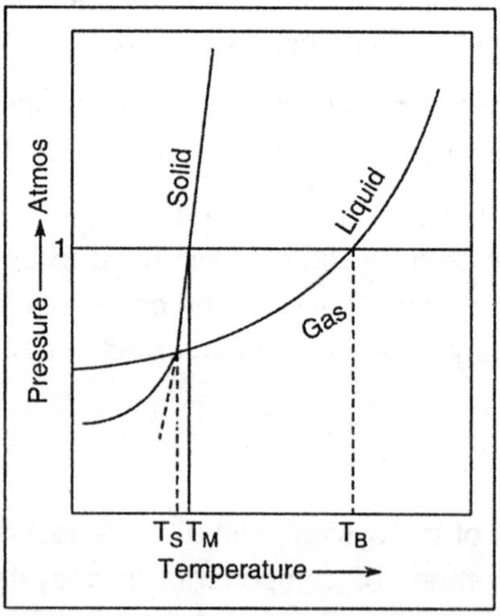

Fig. 6.2 Pressure-temperature relationships to indicate stability regions of solid, liquid and vapour phase in a single component system (shown by solid lines) along-with the metastability regions (shown by dotted lines).

The same conclusion can be drawn from the G-T relationships for solid-liquid-gas phases for a single component system. The free energy values are calculated for all phases as they vary with temperatures and at different pressures. The minimum value of G indicates the stability of that phase under the conditions under consideration. This is shown in Fig. 6.3. It clearly shows that at low pressures vapour is more stable than liquid and hence direct sublimation rather than melting takes place. At high pressures the temperature rise shall result in solid melting to liquid and then converting into vapour. The situation is clarified further by plotting G-T diagram as shown in Fig. 6.4.

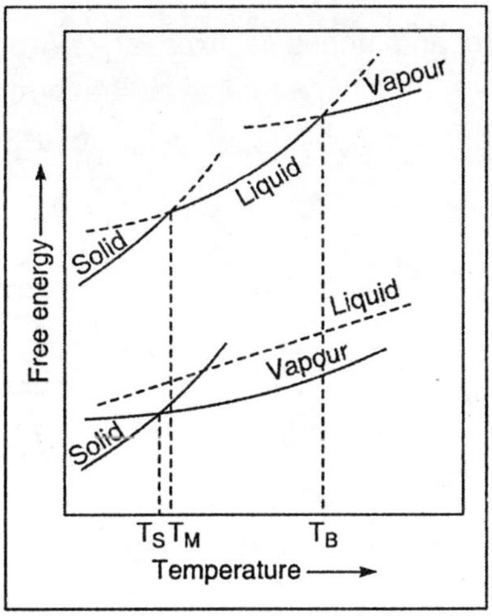

Fig. 6.3 The free energy-temperature relationship to show relative stabilities of phases (shown by solid lines) and their metastable regions (shown by dotted lines).

Below melting point T_M stable phase is solid and above it liquid. The unstable phases are shown by dotted lines. Similarly, if the ΔG of solid to liquid phase change is plotted against temperature the situation is shown in Fig. 6.5. It clearly indicated that ΔG is negative for this only above T_M temperature indicating its possibility and below T_M it is positive and hence not possible.

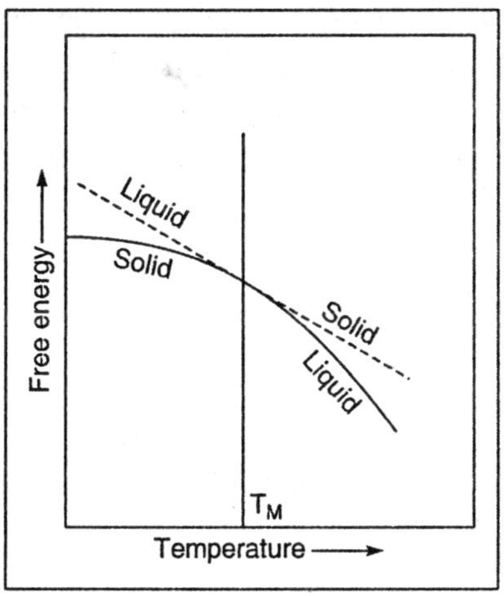

Fig. 6.4 Free energy temperature relationship to indicate phase stability of solid and liquid phases.

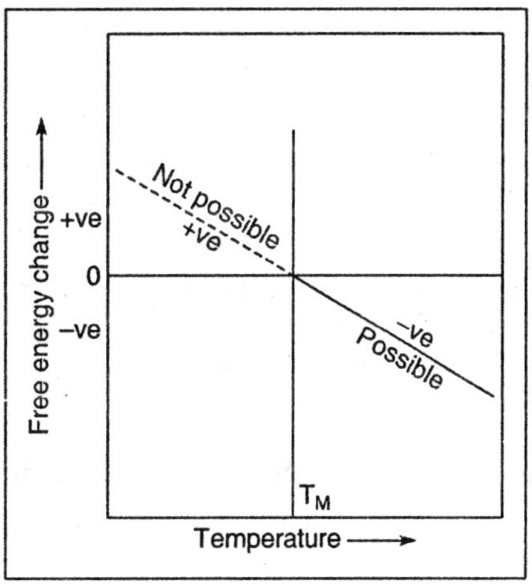

Fig. 6.5 The free energy change as a function of temperature when phase change occurs.

6.5 CONCLUDING REMARKS

Although obvious still the equilibrium between various systems or sub-systems within a system has been established by the zeroth law of thermodynamics. The stabilities of phases can be evaluated knowing their P-T, G-T G-P or G-P-T relationships. This is useful from the point of view of their uses in practice. A phase which is not stable can pose problems if it is to be used in practice over a certain duration.

□□□

Chapter 7

Chemical Equilibria

Equilibrium Constant

7.1 INTRODUCTION

The modern day civilization can sustain itself by converting the naturally occurring materials into a series of products with different chemistry. Even the already changed chemistry is changed further several times to obtain the desired chemistry of the products. Literally therefore, there are innumerable chemical reactions that are carried out in actual practice, particularly in the field of metallurgical and chemical engineering. Although these chemical processes are carried out in such a way that the reaction proceeds in forward direction but the knowledge of equilibrium conditions of such chemical processes is a must to work out the conditions for making the reactions go in the desired direction. It is thus the primary objective of thermodynamic study to first establish the equilibrium conditions of chemical processes of interest. It is from this point of view that this topic has been dealt with independently in this chapter.

7.2 THERMODYNAMIC EQUILIBRIUM CONSTANT

Let us consider a general chemical reaction at constant pressure and temperature as

$$aA + bB + \ldots = cC + dD + \ldots \qquad \ldots(7.1)$$

where capital letters are used to denote the reactants or products and the small letters are used to describe molecules involved in the balanced form of the reaction. The free energy change may be written as

$$\Delta G = \Sigma G \text{ (products)} - \Sigma G \text{ (reactants)}$$
$$= (cG_C + dG_D + \ldots) - (aG_A + bG_B + \ldots) \qquad \ldots(7.2)$$

where individual free energies are described in terms of the respective G values. If the reactants and the products are assumed to exist in their standard states, then the free energy change may be written with suffixes of standard state on each free energy value as

$$\Delta G^\circ = (cG_C^\circ + dG_D^\circ + \ldots) - (aG_A^\circ + bG_B^\circ + \ldots) \qquad \ldots(7.3)$$

Subtracting Eq. (7.3) from (7.2), standard and non-standard state equations of free energy changes and, rearranging the terms suitably, leads to

$$\Delta G - \Delta G^\circ = [c(G_C - G_C^\circ) + d(G_D - G_D^\circ) + \ldots] - [a(G_A - G_A^\circ) + b(G_B - G_B^\circ) + \ldots] \qquad \ldots(7.4)$$

Substituting as derived earlier $G - G° = RT \ln a$ in the above gives

$$\Delta G - \Delta G° = [c\,RT \ln a_C + d\,RT \ln a_D + ...] - [a\,RT \ln a_A + b\,RT \ln a_B + ...] \quad ...(7.5)$$

on rearranging becomes

$$\Delta G - \Delta G° = RT \ln \frac{a_C^c \cdot a_D^d \cdots}{a_A^a \cdot a_B^b \cdots} \quad ...(7.6)$$

if the reaction (7.1) is at equilibrium, *i.e.*, if the reactants and the products are in thermodynamic equilibrium with each other and in that case the parameter, which is of constant value is denoted as :

$$\ln \frac{a_C^c \cdot a_D^d \cdots}{a_A^a \cdot a_B^b \cdots} = K \quad ...(7.7)$$

The term K by definition is called the **thermodynamic equilibrium constant** of the reaction. The activities of individual reactants and products may change but the value of K remains constant at a fixed temperature for a given reaction.

Hence,
$$\Delta G - \Delta G° = RT \ln K \quad ...(7.8)$$

and if the reaction is at equilibrium then the free energy change is zero, *i.e.*, $\Delta G = 0$ and therefore,

$$\Delta G° = -RT \ln K \quad ...(7.9)$$

Putting value of R as 1.987 cal/deg/mole and changing to common logarithm

$$\Delta G° = -4.575\,T \log K \quad ...(7.10)$$

It is thus clear that the free energy is a function of temperature and hence K is also a function of temperature. This is one of the most commonly used relationship in evaluating the chemical equilibria. This evaluation indicates whether a particular reaction would occur in practice, *i.e.*, whether there exists a tendency in the system to move in one or the other direction or is at equilibrium. This is best explained by two illustrations as follows :

■ **Example 7.1** Will magnesia crucible stay in shape in a furnace operating at 1600°C under a vacuum of 10^{-4} atmosphere? Given is the free energy for the reaction

$$2\{Mg\} + \{O_2\} = 2 <MgO>$$

is
$$\Delta G° = -358700 + 102.6\,T \text{ cal/g.mole}$$

Solution : The value of ΔG at 1600 + 273 = 1873 K is given as

$$\Delta G_{1873°} = -166500 \text{ cal/g mole}$$

since
$$K = \frac{a_{MgO}^2}{P_{Mg}^2 \cdot P_{O_2}}$$

$$\Delta G_{1873°} = -RT \ln K = -4.575 \times 1873 \log \frac{a_{MgO}^2}{p_{Mg}^2 \cdot p_{O_2}}$$

putting $a_{MgO} = 1$ for solid MgO, the above reduces to

$$-166500 = +7600 \log p_{Mg}^2 \cdot p_{O_2} \quad \text{or} \quad \log p_{Mg}^2 \cdot p_{O_2} = -21.91$$

In the ambient atmosphere under vacuum of 10^{-4} atoms. the components existing are magnesium vapour and oxygen gas. Each at best can be equal to the vacuum pressure. The proportion of magnesium vapour and oxygen in the gas phase is not known and there is no alternative but to assume that these are in proportion to $1\,p_{O_2} = 2\,p_{Mg}$ moles and hence after substituting this in the above gives $p_{O_2} + 2\,p_{Mg}$ = vacuum pressure. But if dissociation does not occur then no magnesium will be there in the gas phase and hence only oxygen partial pressure shall be equal to the vacuum pressure. First assuming that dissociation does not occur then – 166500 = + 7600 log p_{O_2} or log p_{O_2} = – 21.90 or $p_{O_2} = 10^{-22}$. Under this situation since the ambient partial pressure of oxygen is more than the equilibrium partial pressure of oxygen the reaction shall have a tendency to move towards formation of MgO, that is magnesia shall stay in shape. If now dissociation is considered to occur then product p_{O_2} shall be equal to 1/3 of the vacuum pressure which will still be less than when the dissociation does not occur and hence chances of dissociation of MgO are quite remote. In other words MgO has equilibrium oxygen potential equal to much less than the vacuum pressure at 1600°C and therefore MgO will not dissociate under the conditions of vacuum and is thus stable under the given conditions, *i.e.*, MgO shall stay in shape.

■ **Example 7.2** Will it be possible to anneal copper ingot at 800°C before rolling under a gas mixture of H_2/H_2O equal to 10^2.

Given $4Cu + O_2 = 2Cu_2O$ $\quad\Delta G° = -79700 + 39.3\,T$ cal/g mole

and $2H_2 + O_2 = 2H_2O$ $\quad\Delta G° = -118000 + 26.75\,T$ cal/g mole

Solution : For a reaction $4Cu + 2H_2O = 2Cu_2O + 2H_2$

Hess's law gives $\Delta G° = -38300 + 12.55\,T$ cal/g mole

or for $2Cu + H_2O = Cu_2O + H_2$ $\quad\Delta G° = -19150 + 6.27\,T$

and $\Delta G^{\circ}_{1073°} = -12423$ cal/g mole

The equilibrium constant of this reaction is

$$K = \frac{a_{Cu_2O} \cdot p_{H_2}}{a^2_{Cu} \cdot p_{H_2O}}$$

and assuming the activities of solid Cu and Cu_2O as unity the value of K reduces to

$$K = p_{H_2}/p_{H_2O}$$

and since $\Delta G = -RT \ln K = -4.575\,T \log K$

putting values of 1800°C it gives

$$-12423 = -4908 \log p_{H_2}/p_{H_2O}$$

or $$\log p_{H_2}/p_{H_2O} = \frac{12423}{4908} = 2.53$$

i.e., $$p_{H_2}/p_{H_2O} = 10^{2.53}$$

This gas mixture is richer in hydrogen than that when the ratio is 10^2. Hence the tendency for the reaction of oxidation of copper is to go towards left. The atmosphere in the ingot annealing furnace is reducing towards copper and hence it will be possible to anneal copper in the given case without the danger of its oxidation.

7.3 TYPES OF REACTIONS

Chemical reactions occur either in a single phase as homogeneous reactions or between two or more phases as heterogeneous reactions. The homogeneous reactions can best be illustrated as :

(a) double decomposition of acid and base in an aqueous solution like

$$HCl + NaOH = NaCl + H_2O \qquad ...(7.11)$$

or as

(b) reaction of two gases like

$$2H_2 + O_2 = 2H_2O \qquad ...(7.12)$$

The equilibrium constant of reaction (7.12) is

$$K = p^2_{H_2O}/p^2_{H_2} \cdot p_{O_2}$$

which is characterised in such a way that for a fixed temperature and partial pressure of oxygen the ratio H_2/H_2O is fixed. In other words the equilibrium constant of such reactions shall be proportional to such ratios or it will be equal to fixed ratio for a fixed temperature and fixed oxygen partial pressures.

The heterogeneous reactions essentially occur between two phases like :

(i) solid-solid $\quad <ZnO> + <C> = Zn + CO$

(ii) solid-liquid $\quad [Mn] + <SiO_2> = [Si] + (MnO)$

(iii) liquid-liquid $\quad [S] + [Mn] + (CaO) = (CaS) + (MnO)$

(iv) solid-gas $\quad <Fe_2O_3> + \{CO\} = <Fe_3O_4> + \{CO_2\}$

(v) liquid-gas $\quad 2[C] + \{O_2\} = 2\{CO\}$

The evaluation of equilibrium constant of such reactions need the activities of reactions and those of the products. For the components in a condensed phase, each component will be in the equilibrium with its own vapour pressure in the contacting gas phase. The vapour pressures shall be extremely low and hence could be equated with their fugacities without involving any significant error. These fugacities, being so low will be substantially constant even over a significant temperature and pressure ranges. Therefore, the activities of the component, in the condensed state are taken as equal to their fugacities in gaseous phase, and shall be constant and hence are taken as unity for usual purposes, considering the components in pure state. If these are not in pure state the activities shall be different from unity. Similarly, the actual values of vapour pressures are generally on the lower side and hence are equated to their fugacities and therefore the **activities are taken in the form of their partial pressures.**

If the component is not in pure form, like the one dissolved in molten iron interacting with components dissolved in molten slag or gas then the knowledge of activities is a must for evaluating the equilibrium constant. Alternatively the value of K can be obtained from the free energy data or knowing K the activity of a component can be worked out when others are known.

7.4 VARIATION OF EQUILIBRIUM CONSTANT WITH TEMPERATURE

Since the Eq. (7.9) implicitly means that the $\Delta G°$ and therefore K depends on the temperature, its variation with temperature will have to be evaluated for establishing the value of K at a temperature when its value at some other temperature is known.

The combined form of first and the second laws of thermodynamics gives variation of G with T in the form of an equation as

$$dG = V\,dP - S\,dT$$

This may be applied to when the reactants and the products both are in their standard states for a generalised reaction as stated earlier in Eq. (7.1) as

$$dG_C^\circ = V_C^\circ\,dP - S_C^\circ\,dT$$

$$dG_D^\circ = V_D^\circ\,dP - S_D^\circ\,dT \qquad \text{and so on}$$

$$dG_A^\circ = V_A^\circ\,dP - S_A^\circ\,dT$$

$$dG_B^\circ = V_B^\circ\,dP - S_B^\circ\,dT \qquad \text{and so on} \qquad ...(7.13)$$

multiplying each one by the corresponding number of atoms or molecules involved in the reaction (as depicted by small letters), the overall process can be described as

$$d\,(cG_C^\circ + dG_D^\circ + ... - aG_A^\circ - bG_B^\circ\,...) = (cV_C^\circ + dV_D^\circ + ... - aV_A^\circ - bV_B^\circ\,...)\,dP$$

$$- (cS_C^\circ + dS_D^\circ + ... - aS_A^\circ - bS_B^\circ\,...)\,dT \qquad ...(7.14)$$

The three expressions in the brackets mean the overall free energy change, volume change and entropy change in the overall process. Therefore,

$$d\,\Delta G^\circ = \Delta V^\circ\,dP - \Delta S^\circ\,dT \qquad ...(7.15)$$

assuming one atmosphere pressure as standard conditions and putting

$$dP = 0 \quad \Delta G^\circ = -\,\Delta S^\circ\,dT$$

from definitional form

$$\Delta G^\circ = \Delta H^\circ - T\,\Delta S^\circ$$

therefore,

$$\Delta S^\circ = \frac{\Delta H^\circ - \Delta G^\circ}{T} \qquad ...(7.16)$$

Putting this value in the above

$$d\,\Delta G^\circ = \frac{\Delta H^\circ - \Delta G^\circ}{T}\,.\,dT \qquad ...(7.17)$$

This is one of the forms of **Gibb's-Helmholtz equation** valid at constant pressure. It can be also written as

$$T\,d\,\Delta G^\circ - \Delta G^\circ\,dT = -\,\Delta H^\circ\,dT$$

dividing by T (the integrating factor being $1/T^2$) gives

$$\frac{d\Delta G^\circ}{T} = -\,\frac{\Delta H^\circ dT}{T^2} \qquad ...(7.18)$$

or

$$\frac{d(\Delta G^\circ/T)}{dT} = -\,\frac{\Delta H^\circ}{T^2} \qquad ...(7.19)$$

or

$$\frac{d(\Delta G^\circ/T)}{d(1/T)} = \Delta H^\circ \qquad ...(7.20)$$

and since

$$\Delta G = -\,RT\ln K,\ \text{thus}\ \frac{\Delta H^\circ}{T} = -\,R\ln K$$

or $$d(\Delta G/T) = -R\,d\ln K \quad \text{or} \quad \frac{R\,d\ln K}{dT} = \frac{\Delta H^{\circ}}{T^2} \qquad \text{...(7.21)}$$

therefore, $$\frac{d\ln K}{dT} = -\frac{\Delta H^{\circ}}{RT^2} \quad \text{or} \quad \frac{d\ln K}{d(1/T)} = -\frac{\Delta H^{\circ}}{R} \qquad \text{...(7.22)}$$

This is known as **van't Hoff equation** relating variation of K with T. The effect of T on K depends on the value of ΔH° for the reaction. If the reaction is endothermic (*i.e.*, ΔH° = +ve) then K increases with increasing temperature and if it is exothermic (*i.e.*, ΔH° = –ve) then K decreases with increasing temperature. Integration of this requires knowledge of variation of H with T and which depends on ΔC_P values of the reaction. if ΔH° is independent of temperature then K varies linearly with inverse of temperature (1/T). With increase in temperature the reaction will move in such a way as to nullify the external constraint as per the Le-Chatelier's principle.

■ **Example 7.3** The equilibrium constants of the reaction

$$\langle NiO \rangle + \{CO\} = \langle Ni \rangle + \{CO_2\}$$

at 900°K is 4.53×10^{-1} and at 1100°K is 3.13×10^{-3}.

Find the enthalpy change of this reaction.

Solution : Using Eq. (7.9) $$\Delta G_{900^{\circ}} = -4.575 \times 900 \times 4.53 \times 10^{-1}$$

$$= -1865 \text{ cal/g/mole}$$

similarly, $$\Delta G_{1100^{\circ}} = -1575 \text{ cal/g/mole}$$

$$\frac{\partial G}{\partial t} = -S$$

Hence $$-S = \frac{-290}{-200} = -1.45 \text{ cal/g mole}$$

Since G = H – TS therefore,

$$H_{900^{\circ}} = -1865 + 900 \times (-1.45)$$

$$= -3170 \text{ cal/g.mole}$$

Enthalpy is generally taken as independent of temperature and hence calculated at any one temperature.

7.4 VARIATION OF EQUILIBRIUM CONSTANT WITH PRESSUE

As per Le-Chatelier's principle the reaction involving one or more gaseous components/phases will only be affected by the variation of pressure and the effect will be such that increasing pressure will tend to move the equilibrium in such a way as to nullify the change in pressure as the external constraint. Any reaction which results in pressure increase by way of formation of gaseous component from condensed phases shall tend to move in forward direction upon decrease in pressure and any reaction which results in pressure decrease by way of formation of condensed phase from out of one or more gaseous reactants shall tend to move in forward direction upon increasing pressure and vice-a-versa. For both the situations, for example for a reaction of dissociation of calcium carbonate as

$$CaCO_3 = CaO + CO_2 \qquad K = \frac{a_{CaO} \cdot P_{CO_2}}{a_{CaCO_3}}$$

with increasing pressure the forward reaction will not be favoured and hence K shall increase or reverse reaction will be favoured and vice-versa. If the pressure is decreased by employing vacuum the process of calcination shall take place readily, that is, at a lower temperature as has earlier been shown in Fig. 2.1. In general therefore if a reaction

$$aA + bB = cC + dD$$

represents the reaction involving gaseous components then

$$K = \frac{p_C^c \cdot p_D^d}{p_A^a \cdot p_B^b}$$

will be independent of overall pressure if

$$c + d - a - b = 0$$

and if this is not so then $K_P = K_X \cdot P^{c+b-a-b}$

where K_X is the equilibrium constant expressed in terms of activities as mole fractions. This is of direct importance in reactions involving condensed and gaseous phases where

$$c + d - a - b \neq 0$$

The following example shall illustrate the use of the more vividly.

For a reaction $2H_2 + O_2 = 2H_2O$

$$K_P = (p_{H_2O}^2)/(p_{H_2}^2 \cdot p_{O_2}) = \frac{N_{H_2O}^2 \cdot P^2}{N_{H_2}^2 \cdot P^2} \cdot \frac{1}{N_{O_2} \cdot P}$$

$$= \left(\frac{N_{H_2O}}{N_{H_2}}\right)^2 \cdot \frac{1}{N_{O_2} \cdot P} = \left(\frac{N_{H_2O}}{N_{H_2}}\right)^2 \cdot \frac{1}{p_{O_2}}$$

or $$p_{O_2} = N_{O_2} \cdot P = \left(\frac{N_{H_2O}}{N_{H_2}}\right)^2 \cdot \frac{1}{K_P} \quad \text{or} \quad K_P \cdot p_{O_2} = \left(\frac{N_{H_2O}}{N_{H_2}}\right)^2$$

Also from Eq. (7.8) $K_P = -\exp^{\Delta G} \cdot -\exp^{RT}$

At a given temperature K_P is therefore known in terms of ΔG of the reaction. Therefore, for a certain partial pressure of oxygen or equilibrium oxidising potential the volume ratio of H_2/H_2O can be obtained.

Similarly, equilibrium CO/CO_2 can also be calculated for reaction like

$$2CO + O_2 = 2CO_2$$

■ **Example 7.4** The free energy of the reaction :

$$\langle CaCO_3 \rangle = \langle CaO \rangle + \{CO_2\}$$

is $$\Delta G = +40250 - 34.4\,T \text{ cal/g mole}$$

What is the highest temperature at which this reaction will not occur in open atmosphere. Also find out the temperature if the reaction was carried out under a vacuum of 10^{-5} atoms.

Solution : The equilibrium constant of this reaction is

$$K = a_{<CaO>} \cdot p_{<CO_2>} / a_{<CaCO_3>}$$

Since both CaO and $CaCO_3$ are pure solid phases their activities are taken as unity and hence

$$K = P_{<CO_2>} = 1\,\text{atmosphere.}$$

Since $\Delta G = -RT \ln K = -4.575\, T \log K = 0 = 40250 - 34.4\, T$

or $34.4\, T = 40250 \quad \text{or} \quad T = 1170°K = 897°C$

It means till nearly 900°C is reached, the pressure of CO_2 shall not exceed one atmosphere and hence calcination will not take place till this temperature (see Section 2.6) in open atmosphere. Under a vacuum of 10^{-5} atoms K will be equal to this pressure and hence

$$40250 - 34.4\, T = -4.575\, T \log 10^{-5}$$

or the required temperature shall be 703°K. The decrease in ambient pressure has reduced the calcination temperature.

7.5 FREE ENERGY - TEMPERATURE RELATIONSHIPS

It has been shown that $G = H - TS$ and H and S both changes with temperature as shown in Section (2.10.2) and (3.6). Therefore, G also must vary with temperature. Since both H and S are not measured in absolute terms but only relatively with some reference state, G also is measurable only in terms of reference state or, only changes are measurable. Hence,

$$\Delta G = \Delta H - T\,\Delta S$$

and if all the reactants and products are in their standard states then

$$\Delta G° = \Delta H° - T\,\Delta S°$$

for a process $\Delta H° = \Sigma H^{\circ}_{(products)} - \Sigma H^{\circ}_{(reactants)}$

and $\Delta S° = \Sigma S^{\circ}_{(products)} - \Sigma S^{\circ}_{(reactants)}$

therefore,

$$\Delta G° = [\Sigma H^{\circ}_{(products)} - \Sigma H^{\circ}_{(reactants)}] - T\,[\Sigma S^{\circ}_{(products)} - \Sigma S^{\circ}_{(reactants)}]$$

$$= \Delta H^{\circ}_{(reaction)} - T\,\Delta S^{\circ}_{(reaction)}$$

The way in which enthalpy and entropy can be evaluated has been dealt with in Section (2.11) and Section (3.12). Therefore, for a process

$$A + B = AB$$

$$\Delta G° = G^{\circ}_{AB} - G^{\circ}_{A} - G^{\circ}_{B}$$

this can be illustrated by an example :

■ **Example 7.5** Calculate the standard free energy of the reaction

$$[FeS] + <CaO> = (FeO) + (CaS)$$

from the following data :

(1) $2\,(Ca) + \{O_2\} = 2\,<CaO>$ $\qquad \Delta G^{\circ}_1 = -307100 - 51.28\, T$

(2) $\{S_2\} + 2\,(Ca) = 2\,<CaS>$ $\qquad \Delta G^{\circ}_2 = -283100 - 42.93\, T$

(3) $\{S_2\} + 2[Fe] = [FeS]$ $\Delta G_3^\circ = -64080 - 19.04\,T$

(4) $2[Fe] + \{O_2\} = 2[FeO]$ $\Delta G_4^\circ = -125700 - 30.69\,T$

(5) $<CaS> = (CaS)$ $\Delta G_5^\circ = +1900 + 9.0\,T$

Solution : The required reaction can be worked out using Hess's law by algebraic summation of the given reactions and in turn it also gives the free energy change of the reaction by the corresponding summation of the free energy values given as :

Add (– reaction 3) + (– reaction 1) + 2 (reaction 5) + (reaction 2) + (reaction 4)

to give the required reaction.

(the –ve sign of reaction means the reaction in reverse)

The free energy change is thus :

$$= -\Delta G_3^\circ - \Delta G_1^\circ + 2\Delta G_5^\circ + \Delta G_2^\circ + \Delta G_4^\circ$$
$$= 380 + 14.70\,T \text{ cal/g.mole}$$

The enthalpy ΔH° of chemical reaction is substantially a constant value without variation with temperature. But ΔS° does change with temperature and also with volume considerably. If these are plotted vs. temperature for a reaction the net effect of H and S can be graphically evaluated to work out the variation of G vs. T. The G vs. T relationships are varied depending upon the nature of the phases involved and volume changes accompanying the process. These shall be discussed later in Chapter 14. As per third law, elements do not have enthalpy or entropy at 0°K and therefore the free energy of compound is taken as free energy of formation of compound from respective elements.

In other words for a reaction

$$A + B = AB \qquad \Delta G^\circ = G_{AB}^\circ - G_A^\circ - G_B^\circ = G_{AB}^\circ$$

since G_A° and G_B° are zero.

In order to simplify the matter G_{AB}° (298°K) are determined using specific heat data and Kirchoff's Law and are plotted in tabular forms so that ΔG° for any other reaction at any higher temperature can be evaluated as follows :

For a reaction $MnO + C = Mn + CO$

$$\Delta G_{298}^\circ = G_{CO_{298^\circ}}^\circ + G_{Mn_{298^\circ}}^\circ - G_{MnO_{298^\circ}}^\circ - G_{C_{298^\circ}}^\circ$$
$$= G_{CO_{298^\circ}}^\circ - G_{Mn_{298^\circ}}^\circ$$

The ΔG_T° value at any higher temperature can be evaluated using the enthalpy and entropy data and Kirchoff's Law.

7.6 CONCLUDING REMARKS

The evaluation of thermodynamic equilibrium constant is a major achievement in understanding the chemical equilibria since its knowledge gives the equilibrium position which can be juxtaposed with the real life situation of the reaction to work out, if any, the desired direction in which the reaction has to be made to move. It also gives the parameters which could be used to make the reaction move in the desired direction. It also gives the limitations of making a reaction move in the desired direction as reaction can move only upto equilibrium, and that too, often not even upto the equilibrium state but much ahead of it.

□□□

Chapter 8

Phase Equilibria in Single Component System

Clausius-Clapeyron Equation

8.1 INTRODUCTION

In practice metals are often heated to high temperatures for their processing in either solid or liquid state. As temperature rises the vapour pressure of metals in gas phase increases and may become substantial. The equilibrium vapour pressures of metals are of direct interest. Firstly if metals are to be kept at high temperatures for a prolonged period it can lead to a substantial metal loss via vapour formation. For example in LD steelmaking process almost one percent iron is lost is vapour form which not only is a net loss but poses pollution hazards. If such losses are to be minimised then conditions will have to be generated to make vaporisation minimum. This requires the knowledge of equilibrium vapour pressure of metals or broadly, the thermodynamics of single component multi-phase system is required over a temperature and pressure ranges of interest. Similarly, for dealing with alloys the thermodynamics of muti-component, multi phase system over a temperature and pressure range shall be of interest.

The melting point and boiling point by implication means those temperatures when transitions solid-liquid and liquid-gas occur when the ambient pressure is one atmosphere. When pressure is different, the m.p. or b.p. do change. The equilibrium solid-liquid, liquid-gas or solid-gas, as a function of temperature and pressure therefore needs to be established. The equilibrium phase, *i.e.,* stable phase under a given set of conditions can be ascertained from the P-T, P-G, G-T or P-G-T diagrams for such pure components as described in the previous chapter. The lines separating the two adjoining phases in such diagrams are univarient, *i.e.*, with one degree of freedom and needs to be expressed mathematically to explain the equilibria involved. These are nothing but the interrelationships of two-phase-single-component system with temperature and pressure.

8.2 EQUILIBRIUM IN SOLID-LIQUID SYSTEM

Let us imagine a pure substance system at its melting point T when both solid and liquid coexists in equilibrium at a pressure P. It also means that their free energies are equal as

$$G_l = G_S$$

Let us imagine that pressure changes by an infinitesimal amount dP and as a result the temperature changes by an amount dT to establish a new equilibrium and in that an infinitesimally small amount of solid or liquid may change over to the other phase. If the associated free energy changes in solid and liquid phases are dG_S and dG_l and if the system is isolated then these two free energy changes must match as

$$dG_l = dG_S \qquad ...(8.1)$$

The vapour pressure of both under the given conditions are expected to be very small and hence their fugacities can be equated with vapour pressures considering the behaviour as like ideal gases. In that case if dP is the change in the equilibrium vapour pressure of this system the combined statement of the first and the second law can be applied to both solid and liquid phases as :

$$dG_l = V_l\, dP - S_l\, dT \qquad ...(8.2)$$

$$dG_S = V_S\, dP - S_S\, dT \qquad ...(8.3)$$

It is also certain therefore that

$$V_l\, dP - S_l\, dT = V_S\, dP - S_S\, dT \qquad ...(8.4)$$

or

$$(V_l - V_S)\, dP = (S_l - S_S)\, dT$$

$$\left(\frac{dP}{dT}\right)_{eq} = \left(\frac{S_l - S_S}{V_l - V_S}\right) = \left(\frac{\Delta S_{(l \to S)}}{\Delta V_{(l \to S)}}\right) \qquad ...(8.5)$$

At equilibrium $\Delta G = 0$ and therefore $\Delta H = T\, \Delta S$

Thus

$$\left(\frac{dP}{dT}\right)_{eq} = \frac{\Delta H}{T\, \Delta V} \qquad ...(8.6)$$

where ΔH is the latent heat of melting per g mole and ΔV is the difference in molar volumes of liquid and solid phases. This is known as **"Clapeyron Equation"** interrelating the variations in temperature and pressure for maintaining the equilibrium between the two phases.

If ΔV_{S-l} is positive, as is normally the case, and if ΔH_{S-l} is a positive quantity as is normal, the $\left(\frac{dP}{dT}\right)_{eq}$ will be a positive quantity, *i.e.*, as pressure increases the transition temperature should also increase. In other words the **equilibrium melting point shall increase with pressure and vice-versa**. However, for ice-water system ΔV_{S-l} is negative and hence $\left(\frac{dP}{dT}\right)_{eq}$ is negative and which is fully exploited for the game of skating on ice wherein the skate pressure increase shall decrease the m.p. of ice and hence help maintain ice without melting for skating purposes, *i.e.*, skating is readily possible.

8.3 EQUILIBRIA RELATED TO LIQUID-VAPOUR SYSTEM

In the previous section solid to liquid change was logically analysed to arrive at some mathematical relations. The same logic can be applied here to analyse the liquid-vapour changer. The same thing need not be repeated here but the equations shall differ only with respect to change from solid to liquid and from liquid to gas phase as follows :

Therefore as in Eqs. (8.2) to (8.4)

$$(V_V - V_l)\, dP = (S_V - S_l)\, dT \qquad ...(8.7)$$

or
$$\frac{dP}{dT} = \frac{\Delta S_{(V \to l)}}{\Delta V_{(V \to l)}}$$

and since
$$V_V >> V_l$$

(difference of liquid and equivalent vapour volumes)

therefore,
$$\Delta V = V_V - V_l = V_V$$

or
$$\left(\frac{dP}{dT}\right)_{eq} = \frac{\Delta H_V}{T\,V_V} \quad ...(8.8)$$

as usually the vapour pressures are very low, it can be assumed to behave ideally then vapour pressure will be V = RT/P as per ideal gas equation and hence

$$\left(\frac{dP_V}{dT}\right)_{eq} = \frac{P_V \cdot \Delta H_V}{RT^2}$$ where P_V is equilibrium vapour pressure.

or
$$\left(\frac{dP_V}{P_V}\right)_{eq} = \frac{\Delta H_V}{RT^2}.dT \quad \text{or} \quad \frac{d\ln P}{dT} = \frac{\Delta H_V}{RT^2} \quad ...(8.9)$$

This equation is one of the most famous and useful equations dealing with vapour pressures and is known as **'Clausius-Clapeyron'** equation. If ΔH_V is assumed to be independent of temperature then on integration it leads to

$$\ln P_V = \frac{\Delta H_V}{RT^2} + I \quad \text{where I is the integration constant} \quad ...(8.10)$$

which correlates the vapour pressure with temperature. Since R, T and ΔH are all positive quantities vapour pressure of any substance increases with temperature and vice-versa.

■ **Example 8.1** The densities of solid and liquid lead at its normal melting point of 327°C are found to be 11.05 and 10.645 g/c.c respectively. The heat of fusion of lead is 1190 cal/g at its melting point. Prove that the m.p. of lead at even high pressures practically remains the same.

Solution : The basis of solving this is to correlate the volume change on melting to derivatives of pressure and temperature.

The gram atomic volumes of solid and liquid lead can be calculated as

$$\text{solid lead} = \frac{\text{at. wt. Pb}}{\text{density}} = \frac{207}{11.005} = 18.809 \text{ cc}$$

and similarly for liquid lead is = 19.445 cc

$$\frac{dP}{dT} = \frac{1190 \times 207}{(327 + 273)(18.8096 - 19.445)} = -6443.5 \text{ cal/0°/cc}$$

changing from 1 cal = 41.193 cc atmos = – 26572 atmos/0°

or
$$dT/dP = 1/26572 = 376 \times 10^{-7} \text{ 0°/atmos}$$

When P = 1 atmos m.p. of lead is 327°C,

when P = 100 then dP = 100 and therefore,

$$dT = 3.7 \times 10^{-3} \text{ °C}$$

Hence the melting point of lead shall increase because of application 100 atmosphere pressure by only 0.0037°C and which is negligibly small and hence it practically remains unchanged.

■ **Example 8.2** Vapour pressure of liquid zinc above its melting point of 420°C is given by :

$$\log p_{Zn} = \frac{6400}{T} + 5.5$$

The heat of fusion of zinc is 1600 cal/g atom, for above the m.p. and it is 1400 cal/g atom, below its m.p. Derive a formula for vapour pressure over solid zinc below the m.p.

Solution : For temperature above m.p. 420°C

$$\frac{dP_{(liq)}}{P} = \frac{\Delta H_V}{RT^2} = d \ln P_{(liq)}$$

and for temperature below its m.p.

$$\frac{dP_{(sol)}}{P} = \frac{\Delta H_S}{RT^2} = d \ln P_{(sol)}$$

at the m.p. of 420°C both are equal and eliminating the natural logarithm

$$\log p_{Zn\,(solid)} = \frac{\Delta H_S}{\Delta H_V} \log p_{Zn\,(liq)}$$

and putting values as given

$$= \frac{5600}{T} + 4.9$$

8.3.1 Vapour Pressure - Temperature Relationships

As ΔH is expressed in terms of ΔC_P values and since C_P is expressed in a generalised form as :

$$C_P = a + bT + cT^2 \text{ cal/mole/0°}$$

the enthalpy of vaporisation is expressed as :

$$\Delta H_V = \Delta H_o + \alpha T + \beta T^2 + \gamma T^3 \quad ...(8.11)$$

where ΔH_O, α, β, γ, etc. are all constants for a given substance. This leads to

$$\ln P = -\frac{\Delta H_o}{RT} + \frac{\alpha}{\beta} \ln T + \frac{\beta}{\alpha} T + \frac{\gamma}{2R} T^2 + C \quad ...(8.12)$$

or in a more simplified form

$$\ln P = \frac{A}{T} + B \log T + C \quad ...(8.13)$$

If log p are plotted vs inverse of temperature then straight line plots over a certain temperature range as shown in Fig. 8.1 are observed experimentally. If such an experimental data are available ΔH_V for substances can be calculated by evaluating the slope of these plots.

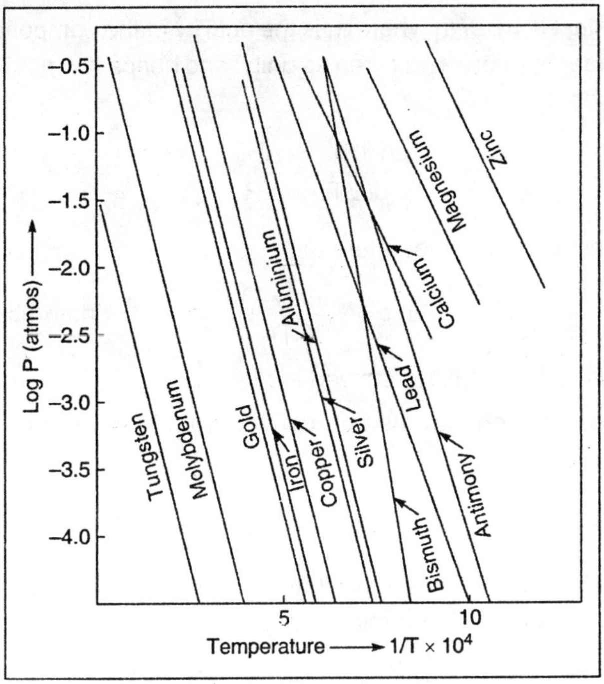

Fig. 8.1 The variation of vapour pressure with temperature as log p vs. 1/T. The near parallel lines and inverse slope are characteristics of this diagram.

8.4 EQUILIBRIA OF SUBLIMATION

The above equations can be suitably modified to develop the equation for the process of sublimation, *i.e.*, the solid-vapour equilibria as :

$$\left(\frac{dP_V}{dT}\right)_{eq} = \frac{P_V \cdot \Delta H_S}{RT^2} \qquad ...(8.14)$$

where ΔH_S is the enthalpy change of the sublimation process.

or
$$\left(\frac{d \ln P_V}{dT}\right) = \frac{\Delta H_S}{RT^2} \qquad ...(8.15)$$

or
$$\ln P_V = \frac{\Delta H_S}{RT^2} + C \quad \text{(integration constant)} \qquad ...(8.16)$$

At the b.p. or the sublimation point the value of P_V is one atmosphere and if ΔH_S value is known then the integration constant can be evaluated and it is used for other calculations thereafter.

8.5 VAN'T HOFF EQUATION AND EQUILIBRIA

The equilibria so far studied and the equations developed to establish the equilibrium relationships of vapour pressure with temperature can also be derived from well known van't Hoff's equation as follows :

The equilibria under study can be described as :

$$\text{liquid} \rightarrow \text{vapour (gas phase)}$$

and the equilibrium constant $K = p_V / a_i$ where a_i is the activity of the component in liquid phase and since it is pure liquid phase therefore a_i is taken as unity, and hence $K = p_V$. Putting this value in the van't Hoff's equation gives

$$\frac{d \ln p_V}{d(1/T)} = \frac{\Delta H_V^\circ}{R}$$

assuming ΔH_V° to be constant with temperature then

$$\ln p = -\frac{\Delta H_V}{RT} + I \qquad \text{(integration constant) as before}$$

this is the same equation as in Eq. (8.10) derived earlier.

Similarly, the same equation can also be derived from Maxwell's relations given earlier in Eq. (4.43).

8.6 TROUTON'S RULE

It has been observed from experimental data that the values of ΔH_V varies linearly with boiling points and that the plots obey broadly the relationship

$$\frac{\Delta H_V}{T_{B.P.}} = 21 \text{ cal/mole/0°} \qquad \ldots(8.17)$$

This is shown in Fig. 8.2. This was first noted by Trouton and hence is known as **'Trouton's Rule'** stating that the entropy of vaporisation of any liquid is approximately equal to 21 cal/mole/0°K. Therefore, at the boiling point this is the measure of increase in randomness on vaporisation and which is approximately the same for many substances including both the metals and their compounds.

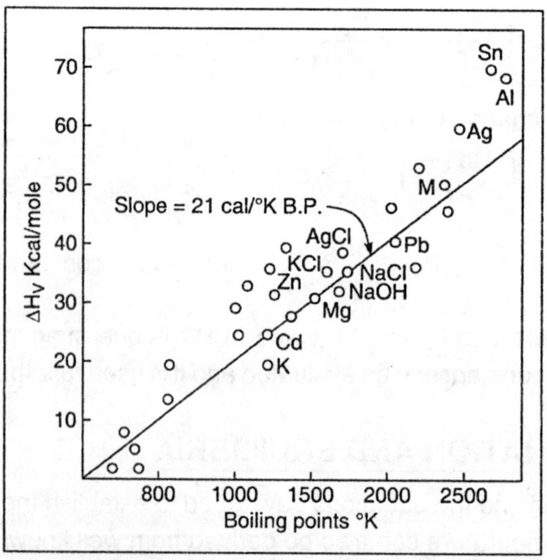

Fig. 8.2 Variation of latent heats of vaporisation of substances as a function of their normal boiling points to indicate the validity of Trouton's Rule.

This is useful because the value of ΔH_V can be evaluated for a substance approximately at the boiling point using this rule. For example at $P_V = 1$, $\Delta H_V = 21 \times T_{B.P.}$ and therefore on substituting these values in Eq. (8.16) gives the integration constant C = 10.57. Using the value of C now vapour pressure of any liquid can be calculated at any other temperature.

■ **Example 8.3** The normal boiling point of copper is 2868°K. Calculate the vapour pressure of copper at 1100°K. What are the assumptions made in solving this?

Solution : The vapour pressure is one atmosphere at normal boiling point. Hence

$$\ln 1 = \frac{\Delta H_V}{1.987 \times 2868} + C$$

assuming that the system obeys Trouton's rule

$$\Delta H_V = 21 \times 2868 \text{ cal}$$

The value of integration constant is thus

$$C = 10568$$

The vapour pressure at 1100°K will be

$$\log p_v = \frac{21 \times 2868}{1.987 \times 1100} + 10.568 \quad \text{or} \quad p_v = 4.6 \times 10^{-7} \text{ atmospheres.}$$

8.7 RAMSAY - YOUNG RULE

It has been observed that for two chemically similar liquids like same family of organic groups or inorganic liquids like oxides, sulphides, etc. there exists an approximate relationship of vapour pressure and temperature. This can best be illustrated as follows :

Let two such similar liquids 1 and 2 have same vapour pressures at temperatures T_1 and T_2 respectively and therefore,

$$\ln p_1 = \ln p_2$$

or

$$\frac{\Delta H_{V(1)}}{RT_1} + C_1 = \frac{\Delta H_{V(2)}}{RT_2} + C_2$$

or

$$\frac{T_1}{T_2} = \frac{\Delta H_{V(1)}}{\Delta H_{V(2)}} - (C_1 - C_2)\frac{RT_1}{\Delta H_{V(2)}}$$

similarly if T_1' and T_2' are another such two temperatures at which the vapour pressures of both are P_V' and hence it reduces to

$$\frac{T_1'}{T_2'} = \frac{\Delta H_{V(1)}}{\Delta H_{V(2)}} - (C_1 - C_2)\frac{RT_1'}{\Delta H_{V(2)}}$$

subtracting one from the other gives

$$\frac{T_1}{T_2} - \frac{T_1'}{T_2'} = \frac{(C_1 - C_2)}{\Delta H_{V(2)}}(T_1 - T_2')$$

experimentally it has been found that the right hand side is almost zero and therefore

$$\frac{T_1}{T_2} = \frac{T_1'}{T_2'} \qquad \text{...(8.18)}$$

or **the ratio of temperature at which the chemically similar liquids have the same vapour pressure is a constant.** This is known as **'Ramsay - Young Rule'**. This is also referred to as **'Duhring's Rule'**. Conversely **'the ratio of boiling points of two chemically similar liquids should have the same value'.**

8.8 RICHARD'S RULE

It was observed from experimental data that the heat of fusion for various substances varied approximately linearly with their melting points. The slope was observed as :

$$\frac{\Delta H_M}{T_{M.P.}} = 2 - 4 \text{ cal/mole/0°} \qquad \text{...(8.19)}$$

This is shown in Fig. 8.3. This is equal to entropy change on melting and since entropy is a measure of randomness of materials, this randomness for various materials on melting is almost the same. This should be obvious.

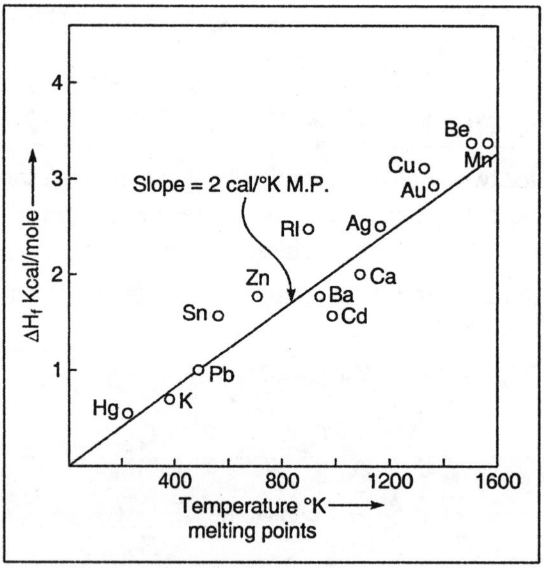

Fig. 8.3 Variation of latent heats of melting of various substances as a function of their melting points to show the validity of Richard's Rule.

8.9 EQUILIBRIUM PHASE TRANSFORMATIONS IN SOLID STATE

The equations derived so far for equilibria between solid, liquid and gas phases can be suitably modified for change in phase in solid state like austenite-martensite transformations. These phases shall have their own vapour pressures in equilibrium in the gas phase. If the solid phases are designated by numbers 1 and 2 then :

$$\left(\frac{dP}{dT}\right)_{eq} = \frac{S_2 - S_1}{V_2 - V_1} = \frac{\Delta S_{(2-1)}}{\Delta V_{(2-1)}} \qquad \text{...(8.20)}$$

or
$$\left(\frac{dP}{dT}\right)_{eq} = \frac{\Delta H_{(1-2)}}{T\,\Delta V_{(2-1)}}$$

where $\Delta H_{(1-2)}$ is the enthalpy of transformation and $\Delta V_{(2-1)}$ is the molar volume change accompanied in the change.

8.10 VARIATION OF VAPOUR PRESSURE WITH TOTAL PRESSURE

Let us imagine a liquid in equilibrium with its vapour at ambient pressure P at a temperature T. If P is one atmosphere then P_V shall be the true vapour pressure of the liquid at temperature T. Let us imagine that the system is flooded with an inert gas to increase the total pressure of the gas phase in equilibrium with the liquid. What will be the vapour pressure of the liquid at the same temperature? Will it be less or more? It is bound to increase with increase in total pressure and decrease if the total pressure decreases, like application of vacuum as per the Le-Chateliar's principle.

Let us therefore, imagine that the system liquid and vapour are at equilibrium at temperature T and pressure P when p is the true vapour pressure. Let us also imagine that now the total pressure increases by dP and correspondingly the vapour pressure increases by dp_V.

Using the combined statement of the first and the second laws the change in the free energy of each of the condensed and the gaseous phases as a result of this infinitesimal change is given by as follows :

for gas phase $\quad dG_g = V_g\, dP - S_g\, dT$

and for condensed phase $\quad dG_C = V_C\, dP - S_C\, dT$

considering isothermal conditions

$dG_g = V_g \,.\, dP_V$ where P_V is vapour pressure

and $\quad dG_g = V_C\, dP_T$ where P_T is the total pressure

In an isolated system like this therefore the two free energy changes shall be equal and hence it reduces to

$$\frac{dP_V}{dP_T} = \frac{V_C}{V_g} \qquad \text{...(8.21)}$$

since V_g is quite small and can be assumed to behave ideally, therefore,

$$V_g = RT/P_V$$

or putting the value of V_g gives
$$\frac{dP_V}{dP_T} = \frac{V_C P_V}{RT}$$

on rearranging
$$\frac{dP_V}{P_T} = \frac{V_C}{RT}\, dP_T \quad \text{or} \quad d \ln P_V = \frac{V_C}{RT}\, dP_T \qquad \text{...(8.22)}$$

or
$$\Delta \ln P_V = \frac{V_C}{RT}\, \Delta\, dP_T \qquad \text{...(8.23)}$$

thus as the total pressure increases the vapour pressure also increases and what was qualitatively suggested earlier is proved here. In the Eq. (8.22) above the $\Delta \ln P_V$ is the difference in vapour pressure at any total pressure P_T and that at any other reference state (normally one atmos) on log scale and ΔP_T is the difference between total pressure and reference state pressure (one atmos).

■ **Example 8.4** Calculate the vapour pressure of zinc under 10 atmos pressure at 700°K. The normal b.p. of zinc is 900°C and density of zinc is 6.5 g/cc. At wt. of zinc is 65.

Solution : Assuming Trouton's rule

$$\Delta H_S \text{ (zinc)} = (900 + 273) \times 21 = 24633 \text{ cal/g.mole}$$

taking $\ln P = -\Delta H_S/RT + 1057$ (where 10.57 is the integration constant)

putting the values $\ln P = 12.4 \times 10^{-2}$ atmos.

This is true vapour pressure and hence

$$\ln p_v = 12.4 \times 10^{-2} \text{ atmos.}$$

Molar volume of zinc is

$$65/6.5 = 10 \text{ cc} = 1 \times 10^{-2} \text{ litres}$$

$$\Delta P_T = P_T - P_V = 10 - 12.4 \times 10^{-2}$$

Putting values in Eq. (8.22) gives

$$\Delta \ln P_V = \left(\frac{1 \times 10^{-2} \, (10 - 12.4 \times 10^{-2})}{0.001616 \text{ or } 700} \right)$$

$$= 0.001616 \quad \text{or} \quad 16.16 \times 10$$

Since $\Delta \ln P_V = \ln P_V - \ln P$ or $\ln P_V = \Delta \ln P_V + \ln P$

or
$$\ln P_V = \left(\frac{1 \times 10^{-2} \, (10 - 12.4 \times 10^{-2})}{0.082 \times 700} \right) + (12.4 \times 10^{-2}) = 0.1256$$

or $P_V = 1.13$ atmos under a total pressure of 10 atmosphere at 700°K.

8.11 CONCLUDING REMARKS

It is seen that it is easier to develop mathematical equations to establish the equilibrium relationships of phase equilibria in a single component multi-phase system and from these to evaluate the vapour pressures of substances as desired. The equilibria in multi-component systems have not been derived to avoid elaboration and complexity.

□□□

Chapter 9

Solutions - I

Gibb's-Duhem Equation and Activity Laws

9.1 INTRODUCTION

Strictly speaking pure element is only a imaginary concept for it does not exist in reality. Even the so-called *'spec pure'* (spectrographic standards) elements do contain impurities, may be of the order of parts per million or parts per billion or perhaps parts per trillion. But the levels of these impurity contents are such that these can be disregarded for practical utilitarian purposes. However, the fact remains that even the so-called pure elements are nothing but *'mixtures'* and since these are homogeneous single phase materials, are referred to as solutions. In practice therefore we deal with solutions alone, whether liquid, solid or gaseous, maybe dilute or very-very dilute or otherwise.

The simplest possible solution is a gas mixture because all gases together form a single homogeneous gas phase at all concentrations of the components. Formation of liquid solutions from two liquids, or liquid and solid, or gas and liquid and so on are innumerable. Formation of solid solutions are also not at all exceptional and many useful alloys or ceramic materials fall in this category.

The composition of a solution is expressed in weight percentages or, atom or mole percentages or, fractions and, partial pressures, particularly for gaseous solutions.

$$(\text{atom }\%)_i = \frac{\text{number of atoms i}}{\text{total number of atoms in solutions}}$$

$$= \frac{n_i}{n_1 + n_2 + n_3 \ldots + n_i}$$

The individual number of atoms can be worked out by dividing the weight percentages by their respective atomic weights. Hence

$$(\text{atom }\%)_i = \frac{(\text{wt }\%)_i/(\text{at. wt})_i}{(\text{wt }\%)_1/(\text{at. wt})_1 + (\text{wt }\%)_2/(\text{at. wt})_2 + \ldots(\text{wt }\%)_i/(\text{at. wt})_i}$$

The molar percentage can be obtained similarly by taking into account moles in place of atoms and molecular weights in place of atomic weights in the above expression.

A solution may consists of two or more components of which the largest in concentration is referred to as solvent and the smaller ones as solutes. A multicomponent solution can be described as consisting of atoms or molecules of components 1, 2, 3, ..., i as $(n_1 + n_2 + n_3 + \ldots + n_i)$ total atoms or moles. If the solution contains only two components then it is termed binary, if three as ternary and so on. The minor uncontrollable impurities contained therein are disregarded. If a large number of components are these like in iron and steel then it is a multicomponent solution.

According to Dalton's law of partial pressures for a gas mixture

$$p_i = N_i\, P \qquad \text{...(9.1)}$$

where p_i is the partial pressure and N_i atomic fraction of the component in solution and P is the total pressure of the gas mixture. The thermodynamic properties of a solution are most often quite different from those of their components because of interaction of the component atoms/molecules and hence these need to be studied precisely. The thermodynamic properties of components in solution is best expressed in terms of their partial properties as explained earlier in Section (4.2). By taking free energy property for the purpose of illustration it has been shown that

$$\overline{G}_i = \left(\frac{\partial G}{\partial n_i}\right)_{T,\,P,\,n_1,\,n_2,\,n_3,\,...,\,n_{i-1}}$$

and the total property G′ of the solution can be obtained by summation as

$$(G)'_{T,\,P} = \sum_1^i G_i\,.\,n_i \qquad \text{...(9.2)}$$

if

$$N_i = \frac{n_i}{n_1 + n_2 + n_3\ ... = n_i}$$

then molar

$$G = \sum_1^i \overline{G}_i\,.\,N_i = \overline{G}_1 N_1 + \overline{G}_2 N_2 + ... + \overline{G}_i N_i \qquad \text{...(9.3)}$$

Similarly, the other extensive properties of the solution can be expressed as :

Molar volume $$\sum_1^i \overline{V}_i\,.\,N_i = \overline{V}_1 N_1 + \overline{V}_2 N_2 + ...\ \overline{V}_i N_i \qquad \text{...(9.4)}$$

Molar enthalpy $$\sum_1^i \overline{H}_i\,.\,N_i = \overline{H}_1 N_1 + \overline{H}_2 N_2 + ...\ \overline{H}_i N_i \qquad \text{...(9.5)}$$

Molar enthalpy $$\sum_1^i \overline{S}_i\,.\,N_i = \overline{S}_1 N_1 + \overline{S}_2 N_2 + ...\ \overline{S}_i N_i \qquad \text{...(9.6)}$$

Molar enthalpy $$\sum_1^i \overline{A}_i\,.\,N_i = \overline{A}_1 N_1 + \overline{A}_2 N_2 + ...\ \overline{A}_i N_i \qquad \text{...(9.7)}$$

free energy and so on.

9.2 GIBB'S DUHEM EQUATION

The differential equation (4.30), at constant pressure and temperature gets reduced to

$$dG' = \left(\frac{\partial G'}{\partial n_1}\right)_{T,\,P,\,n_2,\,n_3,\,...\,n_i} . dn_1 + \left(\frac{\partial G'}{\partial n_2}\right)_{T,\,P,\,n_1,\,n_3,\,...\,n_i} . dn_2$$

$$+ \left(\frac{\partial G'}{\partial n_3}\right)_{T,\,P,\,n_1,\,n_2,\,...\,n_i} . dn_3 + ... + \left(\frac{\partial G'}{\partial n_i}\right)_{T,\,P,\,n_1,\,n_2,\,...\,n_{i-1}} . dn_i \qquad \text{...(9.8)}$$

$$= \overline{G}_1\, dn_1 + \overline{G}_2\, dn_2 + \overline{G}_2\, dn_3 + \ldots + \overline{G}_i\, dn_i \quad \ldots(9.9)$$

Similarly, the total property of a solution is

$$\overline{G}' = n_1\overline{G}_1 + n_2\overline{G}_2 + n_3\overline{G}_3 + \ldots + n_i\overline{G}_i \quad \ldots(9.10)$$

or $$dG' = (n_1 d\overline{G}_1 + n_2 d\overline{G}_2 + n_3 d\overline{G}_3 + \ldots + n_i d\overline{G}_i) + (\overline{G}_1 dn_1 + \overline{G}_2 dn_2 + \overline{G}_3 dn_3 + \ldots + \overline{G}_i dn_i) \quad \ldots(9.10a)$$

On combining Eq. (9.9) and (9.10a) leads to

$$n_1 d\overline{G}_1 + n_2 d\overline{G}_2 + n_3 d\overline{G}_3 + \ldots + n_i d\overline{G}_i = 0 \quad \ldots(9.11)$$

dividing by total number of atoms/moles in the solution, *i.e.,* $(n_1 + n_2 + n_3 + \ldots + n_i)$ gives

$$N_1 d\overline{G}_1 + N_2 d\overline{G}_2 + N_3 d\overline{G}_3 + \ldots + N_i d\overline{G}_i = 0 \quad \ldots(9.12)$$

This is known as **Gibb's Duhem equation.**

If the solution is of only two components namely 1 and 2 then

$$N_1 d\overline{G}_1 + N_2 d\overline{G}_2 = 0 \quad \ldots(9.13)$$

or $$N_1 d\overline{G}_1 = -N_2 d\overline{G}_2 = -(1 - N_1)\, d\overline{G}_2 = (1 - N_2)\, d\overline{G}_1 \quad \ldots(9.14)$$

By the same treatment Gibb's Duhem equation for all other extensive properties can be evaluated as :

for internal energy $$N_1 d\overline{U}_1 + N_2 d\overline{U}_2 + N_3 d\overline{U}_3 \ldots N_i d\overline{U}_i = 0 \quad \ldots(9.15)$$

for entropy $$N_1 d\overline{S}_1 + N_2 d\overline{S}_2 + N_3 d\overline{S}_3 \ldots N_i d\overline{S}_i = 0 \quad \ldots(9.16)$$

for work function $$N_1 d\overline{A}_1 + N_2 d\overline{A}_2 + N_3 d\overline{A}_3 \ldots N_i d\overline{A}_i = 0 \quad \ldots(9.17)$$

for volume $$N_1 d\overline{V}_1 + N_2 d\overline{V}_2 + N_3 d\overline{V}_3 \ldots N_i d\overline{V}_i = 0 \quad \ldots(9.18)$$

for enthalpy $$N_1 d\overline{H}_1 + N_2 d\overline{H}_2 + N_3 d\overline{H}_3 \ldots N_i d\overline{H}_i = 0 \quad \ldots(9.19)$$

and so on. In fact the Gibb's Duhem equation for binary can be stated for any extensive molar property Z as :

$$N_1 d\overline{Z}_1 + N_2 d\overline{Z}_2 = 0 \quad \ldots(9.20)$$

9.3 DETERMINATION OF PARTIAL MOLAR QUANTITIES FROM MOLAR QUANTITIES OF SOLUTION

The molar general extensive property Z of a two component solution is given as

$$Z = N_1\overline{Z}_1 + N_2\overline{Z}_2 \quad \ldots(9.21)$$

differentiation gives

$$dZ = N_1\overline{Z}_1 + N_2 d\overline{Z}_2 + Z_1 dN_1 + \overline{Z}_2 dN_2 \quad \ldots(9.22)$$

Gibb's Duhem equation for a two component solution as given in Eq. (9.20) when added to this leads to

$$dZ = \overline{Z}_1 dN_1 + \overline{Z}_2 dN_2 \quad \ldots(9.23)$$

Now in a binary solution $$N_1 + N_2 = 1$$

hence $$dN_1 + dN_2 = 0 \quad \text{or} \quad dN_1 = -dN_2$$

multiplying both sides of the equation (9,23) by N_1/dN_2 gives

$$N_1 \frac{dZ}{dN_2} = -N_1\overline{Z}_1 + N_1\overline{Z}_2 \quad ...(9,.24)$$

from equation (9.21) $N_1\overline{Z}_1 = Z - N_2\overline{Z}_2$

substituting this value in Eq. (9.24) gives

$$N_1 \frac{dZ}{dN_2} = -Z + N_2\overline{Z}_2 + N_1\overline{Z}_2 \quad ...(9.25)$$

$$= -Z + \overline{Z}_2 (N_1 + N_2)$$

$$= -Z + \overline{Z}_2 \quad ...(9.26)$$

Putting $N_1 = (1 - N_2)$

$$\overline{Z}_2 = Z + (1 - N_2)\frac{dZ}{dN_2} \quad ...(9.27)$$

and similarly, $$\overline{Z}_1 = Z + (1 - N_1)\frac{dZ}{dN_1} \quad ...(9.28)$$

For a multi-component system if should be

$$\overline{Z}_1 + Z + (1 - N_1)\left(\frac{dZ}{dN_1}\right)_{N_2, N_3, ..., N_i} \quad ...(9.29)$$

If only means that if total molar property of a solution, as it varies with composition of one component, is known, then the partial molar property of other components can be found one by one using the above relationship.

9.3.1 Graphical Method

The variation of molar extensive property with composition of a solution is shown in Fig. 9.1. Graphically from such a graph the partial molar property of a component 1 and 2 can be obtained by

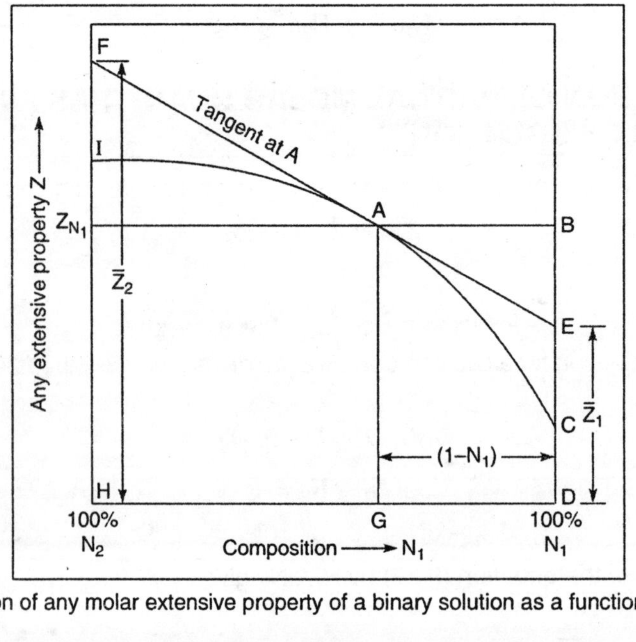

Fig. 9.1 Variation of any molar extensive property of a binary solution as a function of composition.

drawing a tangent on the curve at the relevant point indicating the composition at which the partial property is required. The tangent is shown in Fig. 9.1 at point A when the concentration is N_1. It strikes the Z axis at E at 100% N_1 and at F at 100% N_2 composition. The intercepts at E and F give the values of the corresponding partial properties. The partial properties are should as $\overline{Z}_1$ and $\overline{Z}_2$ as DE and HF respectively.

Graphically, $$\overline{Z}_1 = DE = DB - BE = GA - BE$$

$$GA = Z \text{ as per the graph}$$

and $$BE = BA\frac{BE}{BA} = (1-N_1) \text{ slope of the graph at A}$$

$$= (1-N_1)\cdot\frac{dZ}{dN_1}$$

Now, $$\overline{Z}_1 = DE = Z-(1-N_1)\frac{dZ}{dN_1} \qquad ...(9.30)$$

and similarly, $$\overline{Z}_2 = HF = Z-(1-N_1)\frac{dZ}{dN_2} \qquad ...(9.31)$$

Thus it proves that the tangent drawn on the curve at the relevant composition point, the intercept on the ordinate gives the partial molar quantities.

Also at $$N_1 = 1,\ N_2 = 0 \text{ and } Z = -\overline{Z}_1 \qquad ...(9.32)$$

or it is the same point where curve cuts the ordinate, *i.e.*, at C.

Similarly at $$N_2 = 1,\ N_1 = 0 \text{ and } Z = \overline{Z}_2 \text{ at point J} \qquad ...(9.33)$$

9.3.2 Additional Relationships of Partial Properties

For a solution of large amount $$G' = H' - TS' \qquad ...(9.34)$$

on differentiation

$$\left(\frac{\partial G'}{\partial n_1}\right)_{T,P,n_2,n_3,...,n_i} . dn_1 = \left(\frac{\partial H'}{\partial n_1}\right)_{T,P,n_2,n_3,...,n_i} - T\left(\frac{\partial S'}{\partial n_1}\right)_{T,P,n_2,n_3,...,n_i} \qquad ...(9.35)$$

It means that $$\overline{G}_1 = \overline{H}_1 - T\overline{S}_1 \qquad ...(9.36)$$

It also means that the relationships valid for free energy of a solution is also valid, in the same format, for individual components in the form of the relevant partial molar properties. Similarly, therefore,

$$\overline{H}_1 = \overline{U}_1 - P\overline{V}_1 \qquad ...(9.37)$$

or $$\overline{A}_1 = \overline{U}_1 - T\overline{S}_1 \qquad ...(9.38)$$

In the same way the combined statement of the first and the second laws of thermodynamics for the solution can be expressed as

$$dG' = V'dP - S'dT \qquad ...(9.39)$$

or similarly $$\overline{dG} = \overline{V}_i dP - \overline{S}_i dT \qquad ...(9.40)$$

and all such relationships shall be valid for solution as well like

$$\left(\frac{\partial G'}{\partial P}\right)_{T,\, n_1,\, n_2,\, \ldots,\, n_i} = V' \qquad \ldots(9.41)$$

and

$$\left(\frac{\partial G'}{\partial T}\right)_{T,\, n_1,\, n_2,\, \ldots,\, n_i} = -S' \qquad \ldots(9.42)$$

Also

$$\left(\frac{\partial \overline{G}'_i}{\partial P}\right)_{T,\, n_1,\, n_2,\, \ldots,\, n_i} = -\overline{V}_i \qquad \ldots(9.43)$$

and

$$\left(\frac{\partial \overline{G}'_i}{\partial T}\right)_{T,\, n_1,\, n_2,\, \ldots,\, n_i} = -\overline{S}_i \qquad \ldots(9.44)$$

Similarly since

$$\left[\frac{\partial(G/T)}{\partial(1/T)}\right]_P = H \quad \text{thus} \quad \left[\frac{\partial(G_i/T)}{\partial(1/T)}\right]_P = \overline{H}_i \qquad \ldots(9.45)$$

9.4 IDEAL SOLUTION

9.4.1 Raoult's Law

In the early days the study of solutions was limited to the measurements of vapour pressures of solutions at various concentrations of the dissolved component in binary solution systems. These measurements were readily feasible by actual experimental work. The observations so made were codified by Raoult in the form of a statement, limited to binary solutions only. **"The lowering of the vapour pressure of a solvent due to the addition of solute is equal to the mole fraction of the solute in the solution,"** suggested Raoult.

If p_1 and p_2 are partial pressures of components 1 and 2 in a binary solution at a certain temperature and if p_1° and p_2° are the vapour pressure of pure components 1 and 2 then

$$\frac{P_1^\circ - P_1}{P_1^\circ} = N_2 \quad \text{or} \quad 1 - \frac{P_1}{P_1^\circ} = N_2 \quad \text{or} \quad \frac{P_1}{P_1^\circ} = 1 - N_2 = N_1 \qquad \ldots(9.46)$$

and

$$\frac{P_2^\circ - P_2}{P_2^\circ} = N_1 \quad \text{or} \quad 1 - \frac{P_2}{P_2^\circ} = N_1 \quad \text{or} \quad \frac{P_2}{P_2^\circ} = 1 - N_1 = N_2 \qquad \ldots(9.47)$$

where N_1 and N_2 are the mole fractions of the components 1 and 2 respectively and that $N_1 + N_2 = 1$. In a generalised way therefore

$$N_i = -\frac{P_i}{P_i^\circ} \quad \text{or} \quad \text{since } p_i^\circ \text{ is fixed at a certain}$$

temperature and pressure

$$p_i \propto N_i \qquad \ldots(9.48)$$

or vapour pressure is directly proportional to its mole fraction. If the vapour pressures are low, these can be approximately equated with respective fugacities and hence

$$N_i = \frac{P_i}{P_i^\circ} = \frac{f_i}{f_i^\circ} = a_i \qquad \ldots(9.49)$$

This mathematical statement is known as **Raoult's Law** and is often stated in the form that **'activity of a solute in a solution is equal to its atom or mole fraction'**. This is shown in Fig. 9.2 by the diagonal line as the Raoult's law line. The activity of a component is equal to its actual atom/mole fraction in the ideal solution and hence Raoultian behaviour is considered as an ideal solution behaviour.

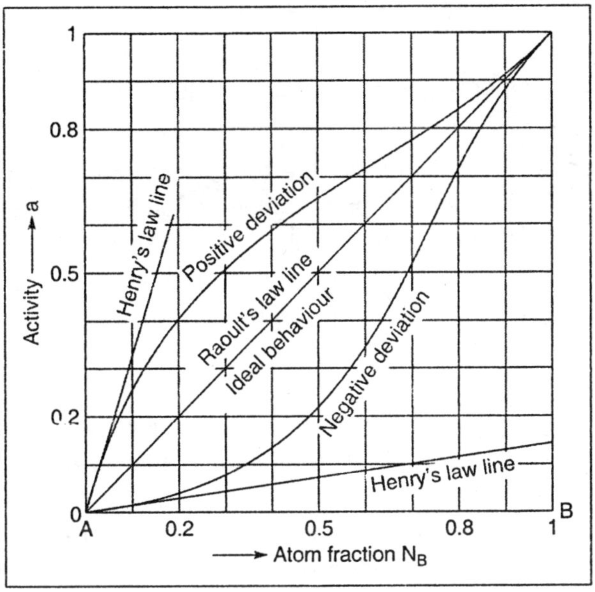

Fig. 9.2 The activity-composition relationship to show Raoultian and Henrian behaviour. The deviation from ideality is also shown.

Those solutions which obey Raoult's law are therefore known as ideal solutions and this holds true at all temperatures and pressures at which the solution is capable of existing. The solution tends to behave ideally at low vapour pressures because the vapour then tends to behave as an ideal gas.

In reality however the solutions hardly obey Raoult's law over the entire composition range Raoult's law is seen to hold true only if the concentration of the solvent is too large and obviously that of the solvent too small. In other words if holds true in dilute solutions only. In non-dilute solutions the behaviour deviates considerably from the ideality in both positive and negative direction. In the deviated form the activity is either more than (+ve deviation) or less than (–ve deviation) the ideal values. The negative deviation is shown in practice by many solutions like Fe-Si, Fe-P, Fe-S, Cu-Sn, Al-Zn, etc. The positive deviation is shown by relatively fewer systems and is typically illustrated by the Fe-Cu system.

Near ideal behaviour is shown by Fe-Mn, Fe-Ni, Fe-Co, ionic melts containing common ion like FeO-MnO, Fe_2SiO_4-$MgSiO_4$ and the like.

9.4.2 Process of Mixing and Component Interaction

In pure state atoms are surrounded on all its sides by like atoms. The process of mixing of components changes the atomic/molecular configuration such that now unlike atoms are available to surround it and such number of unlike atoms surrounding the atom increases on increasing the concentration of the solute in the solution. This gives rise to short range as well as long range ordering even in liquid solutions in certain preferential way. This short or long range ordering basically arises due to different atomic structures of the component atmos and it manifests as interaction of the components in solution. The solution formation is a natural process at least to some extent of concentrations of the components. Hence the components have natural tendency to form short or long range ordering between unlike atoms.

Almost the entire volume together of the components is available for component atoms/molecules to move. In other words the **entropy of the components on solution formation positively increases.**

All this leads to change in the energies of the constituent particles on forming a solution and hence energies will generally be liberated or absorbed depending upon the above changes with reference to the characteristics of the component atoms involved in forming the solution. These thermodynamic changes by no means are insignificant and **treating the solution simply as a mechanical mixture of the components can lead to very erroneous deductions.**

9.4.3 Solution Models

In a solution the activity of a component should ideally be reduced down to its own proportion, *i.e.,* atom/mole fraction. But the real behaviour is quite different. Solutions exhibit wide ranging variations in the activities of the component which in turn means equally wide ranging thermodynamic properties as has been experimentally observed. There is therefore necessity to explain these observations in terms of some models based on interaction of component atoms/molecules in solution.

Three main theories have been suggested to explain this namely :

(1) the classical thermodynamic model assuming formation of one or more of complexes,

(2) the statistical thermodynamic model assuming free volume for vibration of each particle in a solution, and

(3) the quasi-chemical theory which takes into account the preferential distribution of component particles in terms of bonds in a solution.

Various assumptions have been made in developing these models. Atoms are supposed to have a discrete size, valency and electronegativity. When two or more atoms form a solution, the differences in these three properties between the component atoms in solution give rise to certain effects. The difference in sizes of atoms give rise to size factor effect manifesting as strain energy. The valency difference gives rise to coulombic effect which is something akin to effect of groups of atoms in the periodic table. The electronegativity difference gives rise to chemical bond formation. **All these result in changes in energy on solution formation.**

9.4.4 Quasi-Chemical Explanation of Activities

Without going into the intricacies of these theories the qualitative behaviour of solutions can be readily understood from the quasi-chemical theory of solution. Hence it is described below. The name itself indicates that it is something like but not quite in terms of chemistry and chemical bonds of the components in solution.

In condensed form, liquid or solid, atoms do link to each other through formation of chemical bonds. The quasi-chemical approach relies more on these bonds in working out the energy of solution as distinct from their components. This is a limited approach as the energy of solution shall be in terms of all effects of the two types of atoms coming together and not merely in terms of their chemical bondings. The bondings actually extend over a distance, *i.e.,* long-range ordering will have to be considered. The quasi-chemical theory does not incorporate this complication either. Nevertheless it does qualitatively explain the behaviour of solution well and therefore needs to be understood.

This behaviour can be understood from an illustration. Let us imagine that in a class room where several students are seated and if any one wants to run away he can do so with the same vigour if the rest of the students do not exert any influence on him and stay neutral towards him. If one or more do interact with him and, say, try to prevent him from running away, he may not be able to run away with his own vigour for it is now reduced depending upon the extent of force (bonds) exerted on his escaping tendency. If the other students help him to run away by way of push he will be able to run away with added activity or more than his own. The exact influence of the forces exerted by others on him will be a complex one depending upon the forces exerted by individuals on him and on others of amongst themselves.

In a solution since the component atoms are different they do have forces of attraction or repulsion amongst themselves, because of their different electronic structures. This is known as interaction amongst themselves.

In a binary solution say of components A and B the interactions possible are A-B, A-A and B-B. This can be summed up as interaction in like atoms or unlike atoms. If the attraction in like atoms is more than that in unlike atoms then both components may exhibit positive deviation from ideality. On the contrary if attraction of unlike atoms dominate then both may exhibit negative deviation from ideality. This attraction reduces the escaping tendency of the components to the surroundings and the activity shall be reduced to the extent this attraction dominates and hence negative deviation shall be seen in the activity.

The situation becomes little more complex if a third element say C is added to the binary solution. The earlier interaction of A-A, B-B and A-B are now influenced by the interaction A-C,B-C and C-C and whatever are the net final interactive forces, to that extent, the final ideal behaviour shall be affected and accordingly the positive or the negative behaviour shall be observed. The situation becomes far more complex if more than three components are involved, *i.e.,* in multi-component systems.

The Raoultian behaviour will be exhibited by the components if the atoms are similar in nature so that the interactions of like and unlike atoms are correspondingly almost similar as if in their pure form. Strictly speaking such a situation is non-existant because no two different atoms are alike. However, if they are almost alike, the examples of which are quoted earlier, the behaviour observed shall be practically ideal. Such cases are obviously few.

On the other extreme if two different atoms are so widely different that strong interaction exists and in consequence intermetallic compounds like $A_x B_y$ tends to form. Equally true is that if the unlike atom interaction is too feeble and hence like atom attraction dominates then the solution may exhibit miscibility gasp as is seen typically in silicate melts.

9.4.5 Activity Coefficient

From the foregone it is quite clear that the activity of a component can be equated to its atom/mole fraction if ideal in its behaviour, but if there is a deviation then what is the actual activity at a given concentration?

In ideal solution $$a_i = N_i$$

the non-ideal behaviour is covered mathematically as

$$A_i = N_i \times \gamma \qquad ...(9.50)$$

where γ is the multiplying factor that changes the atom fraction N_i into the real value of activity.

This multiplying factor is known as **'activity coefficient'**. The activity coefficient is not a fixed value, it varies with concentration as will be obvious from Fig. 9.2. Most often dilute solutions are under consideration and wherein the non-ideal behaviour conforms to a straight line behaviour indicating that activity coefficient γ has a fixed value for a given binary solution. This activity coefficient is often referred to as γ_i° to indicate that it is valid for dilute concentrations of solute only and hence is distinctly different from γ for any other higher concentrations.

Activity coefficient has a value of unity when ideal behaviour is observed. For negative deviation, γ will be less thanunity while for positive deviation, γ will be greater than unity.

9.4.6 Henry's Law

The behaviour of a solute at dilute concentrations was found to obey

$$a_i = N_i \times i$$

or

$$a_i \propto N_i \qquad ...(9.51)$$

and was noticed and stated and hence is known after its observer as **'Henry's Law'**. It states that **the activity of a component in dilute solution is proportional to its atom/mole fraction** as distinct from Raoult's Law. In reality it has been observed that for dilute solutions, the solute obeys Henry's law whereas the solvent obeys Raoult's law (being at very high) far too apart from each other and hence negligible interaction occurs to affect Raoultian activity.

9.4.7 Standard States and their Interrelations

By definition of activity itself, it is obvious that it is a numerical fraction to express the interactive ability with respect to some standard interactive ability. The interactive ability is maximum when the material is in its purest form. This pure from can be solid, liquid or gaseous form of that substance, as the case may be under given conditions of temperature and pressure. This is purely arbitrary and without any scientific basis, through perfectly alright. The activity of a component in such a pure form is assumed to be unity and the rest are expressed as its fraction when in impure form. The condition of the component is which form the activity is taken as unity is defined as the **'standard state'** of that component for the purpose of calculation of its activities when in non-standard condition.

From the definition of standard state any other state of convenience, other than the pure form can be arbitrarily taken as standard state. The Henry's law offers two such alternative standard states. One of which is **the infinitely dilute atom-fraction standard state**. This is so chosen such that the Henrian activity approaches the value of atom fraction at infinite dilution. If the Henrian activity is depicted by **'h'** (as distinct from **'a'** used to denote Raoultian activity) then

$$h = N_i \quad \text{as} \quad N_i \to 0 \qquad \ldots(9.52)$$

The Henry's law line as shown in Fig. 9.3 is the ideal line for Henry's law and the value of unit activity is shown by γ_i° on the y-axis. Actually the Raoultian activity

$$a = \gamma_i^{\circ} N_i$$

where γ_i° is the activity coefficient, independent of concentration for dilute solutions. On Henrian scale

$$h_i \propto N_i$$

and hence

$$h_i = \gamma_i^{\circ} . N_i \qquad \ldots(9.53)$$

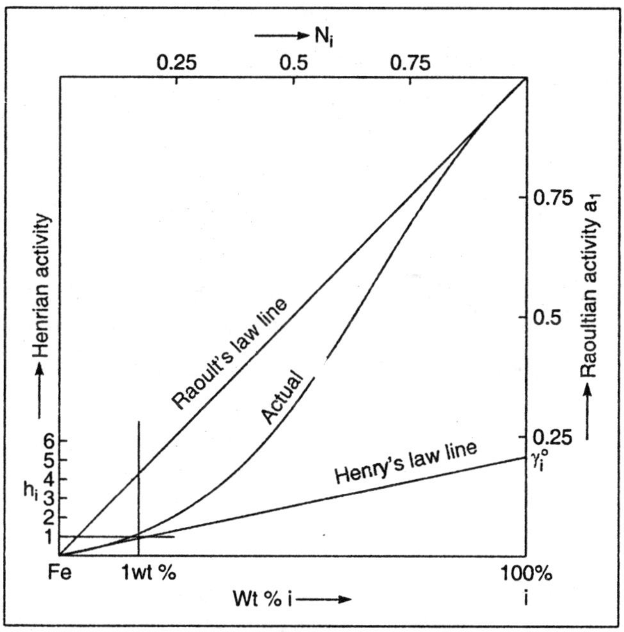

Fig. 9.3 Depiction of Henry's law scale vs. Raoult's law scale to differentiate them.

But γ_i° is the Henry's law coefficient. On assuming the dilute solution standard state for Henry's law

$$h_i = N_i \quad \text{as} \quad \gamma_i^{\circ} = 1$$

But γ_i° is the Henry's law coefficient. On assuming the dilute solution standard state for Henry's law

$$h_i = N_i \quad \text{as} \quad \gamma_i^{\circ} = 1$$

This is one of the standard states on Henrian scale. If there is a deviation from this behaviour then mathematically it is equated as

$$h_i = N_i . f_i$$

where f_i is the activity coefficient on Henrian scale with reference to infinite dilute solution standard state.

Also therefore
$$f_i = \frac{h_i}{N_i} \quad \text{...(9.54)}$$

The interrelationship of these two standard states, namely one as pure substance standard state and the other as infinitely dilute solution standard state can be assessed as

$$\frac{\text{activity of i relative to pure substance standard state}}{\text{activity of i relative to infinitely dilute solution standard state}} = \frac{N_i \gamma_i^\circ}{N_i} = \gamma_i^\circ$$

While dealing with chemical or physico-chemical processes the energy exchange in the process would depend upon the initial state of the system, whether pure, *i.e.,* at unit activity or impure, *i.e.,* at a lower activity. Therefore, it is necessary to know the free energy change involved in transferring one mole of i from one to the other standard state and it is given by

$$\Delta G^\circ = G^\circ_{i\,(\text{pure substance standard state})} - G^\circ_{i\,(\text{dilute solution standard state})}$$

$$\Delta G^\circ = RT \ln \left[\frac{\text{activity of i (pure substance standard state)}}{\text{activity of i (dilute solution standard state)}}\right]$$

$$= RT \ln \gamma_i^\circ \quad \text{...(9.55)}$$

This is still far too complicated to deal with in practice for evaluating activities of energy exchanges.

In a multi-component system the N_i can be evaluated only after complete chemical analysis of the solution to its minute details including the small impurity concentrations. This is rather too difficult. A practical way out is obtainable from Henry's law by way of **assumption of one weight percent concentration of a component as its another standard state**. If this is assumed then 1 wt% concentration will have unit activity on Henrian scale. Strictly speaking it is

$$h_i = \left(\frac{a_i}{a_i/\text{wt\%}_i}\right)_{\text{limit wt\%} \to 0}$$

therefore, at lower concentrations

h_i = wt% i as shown in Fig. 9.3.

The activity values are thus directly the values of chemical analyses of components in dilute ranges obtained as weight percentages. No other components other than the required ones need be analysed in this case unnecessarily.

If the real behaviour deviates from this ideal Henrian behaviour, activity coefficient can be introduced as before as

$$h_i = [\text{wt\% i}] \times f_i$$

or
$$f_i = \frac{h_i}{\text{wt\% i}} \quad \text{...(9.56)}$$

The utility of this practical standard state lies in knowing the free energy associated with the production of 1 wt % dilute solution from a pure component. It is obviously related to the nature of the solvent and will have to be assessed for each solvent separately. For iron solutions it can be shown that the free energy for changing the component from pure state 1 wt% dilute solution in iron, using M_{Fe} and M_i as atomic weights of iron and component i, is given by :

$$\Delta G^\circ = RT \ln \gamma_i^\circ + RT \ln [M_{Fe}/M_i]$$

or

$$= (4.575\ T \log 0.5585\ \gamma_i^\circ)/M_i \qquad ...(9.57)$$

The standard state is like a datum level for measuring the energy exchanges in processes as absolute values are just not measurable. When reactants and products are not in their pure forms any other datum level can be arbitrarily taken for the assessment of energy exchanges. So long as all measurements are made with this new reference state, problems do not arise. But if some measurements are made with one standard state and the others with other standard state then they can be transferred from one to the other standard state by knowing the free energy change for transferring from one to the other standard state.

■ **Example 9.1** Molten silver was contained in a crucible of Si_3N_4 under nitrogen and hydrogen gas mixture at 1400°C for the reaction

$$\langle Si_3N_4 \rangle = 3(Si) + 2\{N_2\}$$

and

$$\Delta G^\circ = 209000 - 96.8\ T \text{ cal/g.mole}$$

At 40% N_2 equilibrium Si content of silver was 0.238%. Calculate the activity coefficient of Si in Ag-Si alloy at 1400°C.

Solution : The free energy change of the above reaction is

$$\Delta G^\circ_{1400} = 209000 - 96.8\ T = 47054 \text{ cal/g/mole}$$

since $K = \dfrac{a_{Si}^3 \cdot P_{N_2}^2}{a_{Si_3N_4}}$ where $P_{N_2} = 0.4$ and $a_{Si_3N_4} = 1$ as solid

therefore,

$$K = (\text{wt \% Si}) \times \gamma\ Si \times 0.4$$

using

$$\Delta G = - RT \ln K$$

$$47054 = - 4.575\ K \{\log ([\text{wt \% Si}] \times \gamma_{Si} \times 0.4)\}$$

putting values in the above gives

$$\log \gamma_{Si} = 7.169 \quad \text{or} \quad \gamma_{Si} = 0.855$$

9.5 ACTIVITIES IN MULTICOMPONENT SOLUTIONS

It has been shown earlier that the activity of a component is affected by the presence of other solute components in the same solution because of their mutual interactions. These interactions are far too complicated for correct predictions but it is necessary that knowing the activities of a component in some simple systems it should be possible to mathematically predict that in a complex solution.

Let the activity coefficient of solute B in a binary solution A-B be given on Henrian scale as

$$f_B = \frac{h_B^{A-B}}{[\text{wt \% B}]}$$

if a third element C is added to this solution then value of f_B^B is altered to the extent C has interaction with A and B and hence

$$h_B^{A-B-C} = h_B^{A-B} \times f_B^C \qquad ...(9.58)$$

where f_B^C is that parameter or multiplying factor that corrects the binary activity into the ternary one. In a way it represents the effect of C on B,

or $$h_B^{A-B-C} = [\text{wt \% B}] \times f_B^{A-B-C} \qquad ...(9.59)$$

where f_B^{A-B-C} is the activity coefficient of B in ternary solution.

It means $$h_B^{A-B-C} = [\text{wt \% B}] \times f_B^B \times f_B^C \qquad ...(9.60)$$

Since f_B itself may vary with concentration it is taken as f_B^B as the actual value at that concentration.

It follows that $$f_B^{A-B-C} = f_B^B \times f_B^C$$

That is the binary activity coefficient when multiplied by the factor, which indicates the influence of C on B, the ternary activity coefficient is obtained. Similarly in a quarternary solution when D is added the activity is given as

$$h_B^{A-B-C} = [\text{wt \% B}] \times f_B^B \times f_B^C \times f_B^D \qquad ...(9.61)$$

provided the influence of D on C in the form of f_B^D is negligibly small to be neglected. This is true in dilute solutions. In a generalised form therefore

$$f_B = f_B^B \times f_B^C \times f_B^D \ldots f_B^i \qquad ...(9.62)$$

for a multi-component dilute solution (A – B – C – D ... i).

Taking logarithms in a multi-component solution

$$\log f_B = \log f_B^B + \log f_B^C + \log f_B^D \ldots + \log f_B^i \qquad ...(9.63)$$

The experimental data on the influence of concentration of third element on the solute, *i.e.,* C or D or i on B in a solvent A, indicate that for dilute solutions the log f_B^i vs wt% of the third element, the influence is typically shown as in Fig. 9.4. When the solution is dilute the relationship is straight line and the slope is given by

$$\text{slope of the line} = \frac{\log f_B^i}{\text{wt\% i}} \qquad ...(9.64)$$

This slope has been designated as interaction coefficient or parameter and is expressed as e_B^C.

Therefore, $$\log f_B^C = e_B^C \times \text{wt \% C} \qquad ...(9.65)$$

where e_B^C indicates the interactive effect of C on concentration of B. Hence

$$h_B = [\text{wt \% B}] \times f_B$$

or

$$= [\text{wt \% B}] \times f_B^B \times f_B^C \times f_B^D \ldots f_B^i \qquad \ldots(9.66)$$

on log scale

$$\log h_B = \log f_B^B + \log f_B^C + \log f_B^D \ldots + \log f_B^i + [\log \text{wt \% B}]$$

$$= e_B^B \,.\, (\text{wt \% B}) + e_B^C \,.\, (\text{wt \% C}) + e_B^D \,.\, (\text{wt \% D}) + \ldots$$

$$e_B^i \,.\, (\text{wt \% i}) + \log \text{wt \% B} \qquad \ldots(9.67)$$

It is therefore possible to experimentally determine activity coefficients in ternary systems and from them compute using the above equation, within reasonable accuracy, the activity of any component in a multi-component system. Although this is valid for dilute solutions the results in the case of even non-dilute solutions have not been all that wrong.

Something e_B^C is known but not e_C^B. This can be found out using Wagner's approximation for dilute solutions as

$$e_B^C M_B = e_C^B M_C \qquad \ldots(9.68)$$

where M indicates the atomic weights.

If both are known their accuracy can be verified using the above equation.

■ **Example 9.2** Calculate the activity of sulphur dissolved in molten iron under the following conditions.

% element in solution	interaction parameter
S – 0.08	$e_S^S = -0.03$
Mn – 0.75	$e_S^{Mn} = -0.025$
Si – 0.30	$e_S^{Si} = +0.065$
P – 0.35	$e_S^P = +0.042$
Cr – 0.67	$e_S^{Cr} = -0.005$
Ni – 1.50	$e_S^{Ni} = -0.002$
C – 2.50	$e_S^C = +0.110$

Solution : The overall activity coefficient of S in the complex iron melt shall be

$$\log f_S = \log f_S^S + \log f_S^{Mn} + \log f_S^{Si} + \log f_S^P + \log f_S^{Cr} + \log f_S^{Ni} + \log f_S^C$$

$$= e_S^S [\% \text{ S}] + e_S^{Mn} \,.\, (\text{wt \% Mn}) + e_S^{Si} \,.\, (\text{wt \% Si}) + e_S^P \,.\, (\text{wt \% P})$$

$$+ e_S^{Cr} \,.\, (\text{wt \% Cr}) + e_S^{Ni} \,.\, (\text{wt \% Ni}) + e_S^C \,.\, (\text{wt \% C})$$

$$= (-0.03 \times 0.08) + (-0.025 \times 0.75) + (0.065 \times 0.30) + (0.042 \times 0.35)$$

$$+ (-0.005 \times 0.67) + (-0.002 \times 1.50) + (0.11 \times 2.50)$$

$$= -0.0024 - 0.00018 + 0.0198 + 0.0147 - 0.0033 - 0.0030 + 0.2750$$
$$= +0.3004$$

Therefore, $f_S = 0.5$

Hence the activity of sulphur dissolved in iron is

$$a_S = \text{wt \% S} \times f_S = 0.08 \times 0.5 = 0.04$$

9.6 SIEVERT'S LAW

Gases dissolve in molten metals readily. The gases like nitrogen, hydrogen, ammonia and oxygen do not dissolve in metals in molecular forms like

$$\{H_2\} = [H_2] \qquad ...(9.69)$$

These however dissolve by dissociation of their molecules as :

$$\{H_2\} = 2[H] \qquad ...(9.70)$$

or

$$\{NH_3\} = [N] + 3[H] \qquad ...(9.71)$$

The dissolution behaviour of diatomic gases like N_2, H_2 and O_2 can be illustrated by the example of hydrogen as shown in Eq. (9.70) above. The equilibrium constant of this reaction shall be

$$K_H = \frac{h_H^2}{f_H} \qquad ...(9.72)$$

Assuming Henrian behaviour and equating fugacity with partial pressures

$$K = K_H = \frac{[\text{wt \% } H_2]}{p_{H_2}} \qquad ...(9.73)$$

or the amount of hydrogen dissolved in the metal is given by as

$$[\text{wt \% H}]^2 \propto (p_{H_2}) \qquad ...(9.74)$$

Similar equations can be developed for oxygen, nitrogen and such other diatomic gases. This was first observed by Sievert and hence is known as **Sievert's Law** which states that '**the solubility of diatomic gases in metals is proportional to the under root of partial pressure of the gas in the ambient atmosphere**' and is stated as

for hydrogen $$[\text{wt \% H}] = K_H (p_{H_2})^{1/2} \qquad ...(9.75)$$

for nitrogen $$[\text{wt \% N}] = K_N (p_{N_2})^{1/2} \qquad ...(9.76)$$

for oxygen $$[\text{wt \% O}] = K_O (p_{O_2})^{1/2} \qquad ...(9.77)$$

It is clear from the above three equations (9.75) to (9.77) that if the metal phase is brought in equilibrium with pure diatomic gas at one atmospheric pressure at a certain temperature then the solubility will be equal to its characteristic constant as

$$[\text{wt \% H}]_{1 \text{ atoms}} = K_H \qquad ...(9.78)$$

Now since solubility increases with temperature these characteristic constants are expressed as a function of temperatures as

$$\log K_N = -\frac{188.1}{T} - 1.246 \qquad ...(9.79)$$

or

$$\log K_H = -\frac{1670}{T} - 1.68 \qquad ...(9.80)$$

and from these the values of characteristic constant, *i.e.,* solubility values under pure gas at one atmosphere pressure at any temperature of interest can be calculated and from this in turn the solubilities under any other conditions can be calculated.

■ **Example 9.3** Molten alloy steel was found to contain10 ppm hydrogen when inside the furnace where the ambient partial pressure of hydrogen was 0.2 atmos. The steel must contain only 1 ppm before it solidities. To what vacuum pressure it must be exposed before casting?

Solution : The Sievert's law gives

$$[\text{wt \% H}]_{steel} = K_H \times p_{H_2}^{1/2}$$

and putting values

$$1 \text{ ppm} = 0.0001 \text{ wt \% } H_2$$

$$0.001 = K_H \times (0.2)^{1/2}$$

or

$$K_H = 2.2 \times 10^{-2}$$

When 1 ppm is the limit

$$0.0001 = 0.022 \times p_{H_2}^{1/2}$$

or

$$p_{H_2}^{1/2} = \left(\frac{0.0001}{0.022}\right)$$

or

$$p = 2.06 \times 10^{-5} \text{ atmos.}$$

The vacuum should be better than 4.5×10^{-4} atmos.

■ **Example 9.4** Find out the amount of nitrogen dissolved in steel melt in a Bessemer process of steelmaking in which air is blown for refining the bath. Later argon was bubbled through the steel both to reduce this nitrogen to 40 ppm for better solidification. Find out the amount of argon bubbled?

Given :

$$\log K_N = -\frac{188}{T} - 1.246$$

Solution : Assuming the temperature of the steel at the end of refining as 1600°C. The value of K_N comes out to be

$$\log K_N = -\frac{188}{1600 + 273} - 1.246 = -1.346$$

% nitrogen dissolved in the steel bath will be that what is in equilibrium with air, *i.e.,* with $p_N = 0.79$.

Therefore, $[\% \text{ N}] = 0.045 \times 0.79^{1/2} = 0.040 = 400 \text{ ppm}$

This is to be reduced down to 40 ppm. It means that 360 ppm are to be reduced, or nitrogen % be brought down from 0.04 to 0.004.

Let us find out the nitrogen partial pressure in a gas phase in equilibrium with 0.004% N_2 in the melt. Hence

$$p_{N_2} = \left(\frac{[\% \text{ N}]}{K_N}\right)^2 = \left(\frac{0.004}{0.045}\right)^2 = 0.29 \text{ atmos.}$$

The balance partial pressure must be that of argon and hence

$$p_A = 1 - 0.29 = 0.71$$

The amount of nitrogen removed per tonne is

$$= 0.00036 \text{ T/ton steel bath}$$

$$= \frac{0.00036 \times 10^3}{28} \times 22.4 \text{ at} = 0.28 \text{ m}^3 \text{ at STP}$$

Equivalent argon volume shall be

$$= 71/29 \times 0.28 = 0.68 \text{ m}^3 \text{ at STP.}$$

Therefore, per ton of steel bath nearly 0.68 m^3 argon gas will have to be bubbled to scavenge nitrogen from the bath provided that every time equilibrium is reached with the argon bubbles.

9.7 CONCLUDING REMARKS

Solutions are by themselves a category of materials possessing their own thermodynamic properties as distinct from those of their constituents. It is because of this reason that a separate branch of solution thermodynamics has been evolved in the later half of the twentieth century to understand the behaviour of solutions. The basic nature of solutions in the form of their models have been dealt with in this chapter and the finer properties shall be dealt with in the next chapter.

□□□

Chapter 10

Solutions - II

Mixing and Excess Functions, and Regular Solution

10.1 INTRODUCTION

If pure silicon and silicon dissolved in iron, under similar conditions, are separately oxidised by identical surroundings, will the free energy change per mole of silicon in both cases be the same? If not why not? The answer is that the two values of ΔG can not be the same. If these are not the same then the free energy associated with the reactants, *i.e.,* silicon in two different forms, pure in one case and impure in the other, must be dissimilar. These two values have got to be different because these are different systems of the same material, under similar conditions of pressure and temperature, and hence associated energy will have to be different. If these were alike very anamolous situation can arise. Fact is that these are not alike. Such situations are met almost always in metallurgical field.

Now therefore, the difference in free energy changes associated with these two are required to evaluate the net free energy change from pure to impure form. The essential difference between these two processes is :

(1) Pure silicon $\rightarrow$ oxidation $\rightarrow$ silicon oxide

(2) Pure silicon $\rightarrow$ dissolved in iron $\rightarrow$ oxidation $\rightarrow$ silicon oxide

The situation shall be complicated further if instead of formation of pure silica, a silicate is formed. In that case the process (2) will be further modified by addition of the step.

silicon oxide $\rightarrow$ reaction with base $\rightarrow$ silicate

at the end. In process (2) free energy change associated with the formation of solution of silicon in iron and that of silica in base matrix are additionally required in assessing the overall change.

The energy associated with pure element at any temperature is merely enthalpy by way of its having been heated above the absolute zero temperature. But as solution the free energy of formation of the solution from the constituents is required. Pure elements do not have any free energy at absolute zero temperature because pure elements do not have any enthalpy and entropy at absolute zero temperature. The free energy of formation of solution,which is often not at all negligibly small, has to be determined. It is often referred to as the free energy of mixing or free energy of formation of solution and is generally expressed by the symbol ΔG^M, with M as the superscript to indicate the mixing nature of the process.

10.2 FREE ENERGY OF MIXING

When to a large amount of solvent or to a solution, if a gram mole of another component is added, the free energy change of this process is the molar free energy of mixing or integral free energy of mixing and is designated as :

$$\Delta G^M = G_{(solution)} - \sum_1^i G_{i\,(pure\ components)}$$

$$= (N_1\overline{G}_1 + N_2\overline{G}_2 + \ldots + N_i\overline{G}_i) - (N_1G_1^\circ + N_2G_2^\circ + \ldots + N_iG_i^\circ)$$

$$= N_1(\overline{G}_1 - G_1^\circ) + N_2(\overline{G}_2 - G_2^\circ) + \ldots + N_i(\overline{G}_i - G_i^\circ) \qquad \ldots(10.1)$$

where $\overline{G}_i$ is the partial molar free energy of component i and G_i° is the molar free energy of the chosen standard form of the i component. The generalised entity $(\overline{G}_i - G_i^\circ)$ is known as partial molar free energy of mixing or relative partial molar free energy and often denoted by $\overline{G}_i^M$ and hence Eq. (10.1) reduces down to molar free energy of mixing as :

$$\Delta G^M = (N_1\overline{G}_1^M + N_2\overline{G}_2^M + \ldots + N_i\overline{G}_i^M) \qquad \ldots(10.2)$$

on the same lines all other extensive properties can be evaluated as :

molar entropy of mixing

$$\Delta S^M = (N_1\overline{S}_1^M + N_2\overline{S}_2^M + \ldots + N_i\overline{S}_i^M) \qquad \ldots(10.3)$$

or molar enthalpy of mixing

$$\Delta H^M = (N_1\overline{H}_1^M + N_2\overline{H}_2^M + \ldots + N_i\overline{H}_i^M) \qquad \ldots(10.4)$$

It can be mathematically proved that as the thermodynamic relationships hold true for pure components, relationships of *similar format* hold true for mixing functions as well :

For total free energy of solution

$$\Delta G^M = \Delta H^M - T\,\Delta S^M \qquad \ldots(10.5)$$

and partial molar free energy of ith component

$$\Delta\overline{G}_i^M = \Delta H_i - T\,\Delta S_i \qquad \ldots(10.6)$$

or

$$\Delta\overline{G}_i^M = RT \ln a_i \qquad \ldots(10.7)$$

The concept of a component in a solution having energy parameter $\overline{G}_i$ different from G_i° has been explained in Chapter 4 in Fig. 4.1. This difference emanates from the fact that the dissimilar atoms interact and alter the short and long range ordering and hence, this **change in bond strength is seen in the form of absorption or release of energy during the solution formation, *i.e.*, the dissolution process.** Obviously the G_i° will have to be different from $\overline{G}_i$. That is why for dissolution reactions like, *e.g.*, silicon in iron

$$\langle Si \rangle = [Si]_{1\,wt\%} \qquad \ldots(10.8)$$

$$\Delta G^\circ = -28000 + 5.54\,T$$

that is

$$\Delta G_{1873^\circ K} = -17620 \text{ cal/g.mole}$$

and for $$\langle C \rangle = [C]_{1\,wt\%} \qquad ...(10.9)$$

$$\Delta G^\circ = +5400 - 10.1\,T$$

and hence $$\Delta G^\circ_{1873^\circ K} = -13500 \text{ cal/g.mole}$$

It only means that the dissolution process would make them more stable in solution form by **liberating energy during the process of dissolution**. It would be interesting to note that although the boiling points of both sulphur and phosphorus are too low, a considerable amount of both of these can remain in solution in molten iron even at such high temperatures as around 1600°C. This is because the free energy changes associated with their dissolution process are quite negative, because of their strong interaction with even molten iron, and hence the observed results.

10.3 IDEAL MIXING OF GASES

Gases form solutions at all concentrations because of free inter-molecular space available to accommodate molecule of other gases. Some metals do form solutions over the entire composition range, both in liquid and solid state, and many over partial composition range. When the attraction between like atoms or molecules dominates over unlike ones the chance of reaching solubility limit at certain concentration may be more.

When two ideal gases mix, because of each being an assemblage of volumeless non-interacting particles, there are no interactions between two ideal gas particles, consequently no bond changes, and hence no heat of mixing or enthalpy of mixing shall be available and hence the enthalpy change on mixing is zero.

In Section (4.4) and Eq. (4.58) it was shown that

$$\overline{G}_i = G_i^\circ + RT \ln p_i$$

and since $p_i = N_i P$ therefore

$$\overline{G}_i = G_i^\circ + RT \ln N_i + RT \ln P \qquad ...(10.10)$$

where P is the total pressure of the gas mixture. Dividing by T and differentiating with respect to temperature gives

$$\frac{\partial(\overline{G}_i/T)}{\partial T} = \frac{\partial(G_i^\circ/T)}{\partial T} \qquad ...(10.11)$$

the latter two terms being independent of temperature vanish.

From Gibb's-Helmholtz equation

$$\frac{\partial(\overline{G}_i/T)}{\partial T} = \frac{\overline{H}_i^\circ}{T^2} \quad \text{and} \quad \frac{\partial(G_i^\circ/T)}{\partial T} = \frac{H_i^\circ}{T^2} \quad \text{or} \quad H_i^\circ = 0 \qquad ...(10.12)$$

In other words the pure phase enthalpy is equal to partial molar enthalpy of the ith component or the enthalpy of the gas mixture is equal to the total enthalpy of unmixed components.

In other words $$\Delta H^M = \sum_1^i mn_i\overline{H}_i - \sum_1^i n_i H_i^\circ = 0 \qquad ...(10.13)$$

or there is no change in enthalpy on mixing as qualitatively and logically indicated above.

The free energy of mixing of ideal gases will have to be a negative value because the process of mixing itself is a natural process. Earlier the particles had freedom to move or randomness within their own pure state volumes, and now have randomness over total volume of the solution (assuming pressure remaining constant), the ΔS is definitely +ve quantity and hence ΔG is –ve quantity.

Free energy of mixing of ideal gases is given by

$$\Delta G^M = G_{(mixture)} - G_{(unmixed\ components)}$$

$$= \sum_1^i n_i \overline{G}_1 - \sum_1^i n_i G_i^\circ = \sum_1^i n_i(\overline{G}_i - G_i^\circ)$$

Putting Eq. (4.58) $\overline{G}_i = G_i^\circ + RT \ln p_i$ $\quad \Delta G^M = \sum_1^i n_i\ RT \ln \frac{p_i}{P}$ ...(10.14)

and if mixing is carried out at constant pressure then

$$\Delta G^M = \sum_1^i n_i\ RT \ln N_i \quad \text{since } N_i = \frac{p_i}{P} \qquad ...(10.15)$$

Therefore, $\quad \Delta G^M_{(ideal\ gases)} = \Delta H^M - T\,\Delta S^M$

and since $\Delta H^M = 0$ thus $\quad = -T\,\Delta S^M$

Also therefore, $\quad \Delta S^M = -\frac{\Delta G^M}{T} = \sum_1^i n_i\ R \ln N_i$ ...(10.16)

$$= \sum_1^i n_i\ R \ln p_i \qquad ...(10.17)$$

If the gases are imperfect then in stead of partial pressures fugacities will have to be taken into consideration and the deviation from above relationships shall be evident.

10.4 IDEAL CONDENSED SOLUTION

If has been shown that for any solution

$$\Delta G^M = G_{(solution)} - G_{(pure\ components)}$$

$$= \sum_1^i (G_i^\circ + RT \ln p_i) - \sum_1^i (G_i^\circ + RT \ln p_i^\circ)$$

$$= \sum_1^i RT \ln \frac{p_i}{p^\circ} = \sum_1^i RT \ln a_i \quad \text{as in Eq. (10.7)}$$

Therefore, $\quad \Delta G^M = (N_1\overline{G}_1 + N_2\overline{G}_2 + \ldots + N_i\overline{G}_i)$

$$= RT\,(N_1 \ln a_1 + N_2 \ln a_2 + \ldots + N_i \ln a_i) \qquad ...(10.18)$$

By definition, for an ideal solution obeying Raoult's law $a_i = N_i$ as the different components do not interact and the Eq. (10.18) reduces to

$$\Delta G^M = RT\,(N_1 \ln N_1 + N_2 \ln N_2 + \ldots + N_i \ln N_i) \qquad ...(10.19)$$

Now, since $-\Delta S^M = \left[\dfrac{\partial \Delta G^M}{\partial T}\right]$, *i.e.*, differentiating the Eq. (10.18) with respect to temperature gives :

$$-\Delta S^M = RT\,(N_1 \ln a_1 + N_2 \ln a_2 + \ldots + N \ln a)$$

$$+ RT\left[N_1\left(\frac{\partial \ln a_1}{\partial T}\right) + N_2\left(\frac{\partial \ln a_2}{\partial T}\right) + \ldots + N_i\left(\frac{\partial \ln a_i}{\partial T}\right)\right] \quad \ldots(10.20)$$

Now since, $\Delta H^M = \Delta G^M + T\,\Delta S^M$

and substituting in this the values from Eq. (10.18) and (10.20) gives

$$\Delta H^M = RT\,(N_1 \ln a_1 + N_2 \ln a_2 + \ldots + N_i \ln a_i) - RT\,(N_1 \ln a_1 + N_2 \ln a_2 + \ldots + N_i \ln a_i)$$

$$- RT_2\left[N_1\left(\frac{\partial \ln a_1}{\partial T}\right) + N_2\left(\frac{\partial \ln a_2}{\partial T}\right) + N_i\left(\frac{\partial \ln a_i}{\partial T}\right)\right] \quad \ldots(10.21)$$

$$= - RT_2\left[N_1\left(\frac{\partial \ln a_1}{\partial T}\right) + N_2\left(\frac{\partial \ln a_2}{\partial T}\right) + \ldots + N_i\left(\frac{\partial \ln a_i}{\partial T}\right)\right] \quad \ldots(10.22)$$

since the first two terms on the right hand side cancel each other.

An ideal solution activity is equal to mole fraction and which is independent of temperature, and hence all the $\left(\dfrac{\partial \ln a_i}{\partial T}\right)$ terms shall be zero and therefore,

$$\Delta H^M_{(ideal)} = 0 \quad \ldots(10.23)$$

as was obtained for ideal gases earlier in Eq. (10.13). Obviously,

$$\Delta S^M_{(ideal)} = - R(N_1 \ln a_1 + N_2 \ln a_2 + \ldots + N_i \ln a_i) \quad \ldots(10.24)$$

$$= - R(N_1 \ln N_1 + N_2 \ln N_2 + \ldots + N_i \ln N_i) \quad \ldots(10.25)$$

and partial molar entropy of mixing in ideal solution is therefore,

$$\Delta \overline{S}^M_{i\,(ideal)} = - R \ln N_i \quad \ldots(10.26)$$

$$\Delta G^M_{(ideal)} = - RT\,(N_1 \ln N_1 + N_2 \ln N_2 + \ldots + N_i \ln N_i) \quad \ldots(10.27)$$

Alternatively the same is proved by Eq. (10.19) and hence partial molar free energy of mixing in ideal solution is therefore,

$$\Delta \overline{G}^M_{i\,(ideal)} = - RT \ln N_i \quad \ldots(10.28)$$

On the same lines if $(G_i - G_i^\circ)$ are differentiated with respect to pressure, which is equal to volume change, then it can be proved that

$$\overline{V}_i = V_i^\circ \quad \text{or} \quad V_i^\circ - \overline{V}_i = 0 \quad \ldots(10.29)$$

that is the volume change on ideal solution formation is zero or

$$\Delta V^M_{(ideal)} = 0 \quad \ldots(10.30)$$

If at this stage Fig. 4.1 is recalled then, if the solution formed is ideal no volume change will be observed.

10.5 REGULAR SOLUTION

The concept of an ideal solution is only a concept and seen to be followed in practice only exceptionally. The actual or real solutions are for out from ideality. While studying the behaviour of a large number of solutions it was observed that although the enthalpies of formations of solutions are widely different their entropy of solution formation are not all that widely spread. In fact in many cases it was observed that the **entropies of formation of solutions are just about the same as if they behaved as ideal solutions.** This is logical enough because the entropy is essentially a configurational property and depends on freedom available for the particles for their movements. When a solution is formed, whether ideal or non-ideal, the freedom available is practically equal to the total volume of solution available and it is almost the same for many solutions, since the volume changes on solution formation being generally negligibly small. This is the reason why $\Delta S^M_{(ideal)}$ is seen to be nearly the same as $\Delta S^M_{(real)}$.

Based on the above observations, **Hildebrand and co-workers** introduced the concept of **'Regular Solution'** as in way between the ideal and the real solutions. By definition it means that **'In regular solution the entropy of formation is equal to as if it was an ideal solution'**. Therefore by definition as derived earlier in Eq. (10.25).

$$\Delta S^M_{(reg)} = \Delta S^M_{(ideal)} = -RT\,(N_1 \ln N_1 + N_2 \ln N_2 + \ldots + N_i \ln N_i) \qquad \ldots(10.31)$$

Hence partial molar entropy of mixing of regular solution is

$$\Delta \overline{S}^M_i = -R \ln N_i \qquad \ldots(10.32)$$

however the $\Delta H^M_{(reg)}$ = whatever finite value the solution has.

Therefore, $\Delta G^M_{(reg)}$ is given as

$$\Delta G^M_{(reg)} = RT\,(N_1 \ln a_1 + N_2 \ln a_2 + \ldots + N_i \ln a_i) \text{ same as in Eq. (10.18).}$$

Hence the partial molar free energy of mixing of a regular solution is

$$\Delta \overline{G}^M_{i(reg)} = \Delta \overline{G}^M_{i(real)} = RT \ln a_i \qquad \ldots(10.33)$$

combining these with the equation giving entropy change gives

$$\Delta H^M = \Delta G^M + T\,\Delta S^M$$

$$= RT\,[(N_1 \ln a_1 + N_2 \ln a_2 + \ldots + N_i \ln a_i) - RT\,(N_1 \ln N_1 + N_2 \ln N_2 + \ldots + N_i \ln N_i)] \qquad \ldots(10.34)$$

and using $a_i = N_i \times \gamma_i$ or $\ln \gamma_i = \ln a_i - \ln N_i$

$$\Delta H^M_{(reg)} = RT\,(N_1 \ln \gamma_1 + N_2 \ln \gamma_2 + \ldots + N_i \ln \gamma_i) \qquad \ldots(10.35)$$

Since this is the total solution property, *i.e.*, integral property, the partial molar property is obviously given by

$$\Delta \overline{H}^M_i = RT \ln \gamma_i \qquad \ldots(10.36)$$

10.5.1 Alpha Function

While analysing the results of activity coefficients of various binary solutions it was found that the factor $\frac{\ln \gamma_2}{(1-N_i)}$ was substantially constant and was therefore, called as α **function**. It can thus be put as

$$\alpha_1 = \frac{\ln \gamma_1}{(1-N_1)^2} \quad \text{and} \quad \alpha_2 = \frac{\ln \gamma_2}{(1-N_2)^2} \qquad \ldots(10.37)$$

for a binary solution or components 1 and 2. It was also observed that $\alpha_1 = \alpha_2$ and hence described as only α and therefore,

$$\alpha = \frac{\ln \gamma_1}{(1-N_1)^2} = \frac{\ln \gamma_2}{(1-N_2)^2} \qquad \ldots(10.38)$$

or

$$\frac{\ln \gamma_2}{\ln \gamma_1} = \frac{N_2^2}{N_1^2} \qquad \ldots(10.39)$$

putting Eq. (10.36) in this gives

$$\frac{\ln \gamma_2}{\ln \gamma_1} = \frac{N_2^2}{N_1^2} = \frac{\Delta \overline{H}_2^M}{\Delta \overline{H}_1^M} \qquad \ldots(10.40)$$

In other words log-activity-coefficients are inversely proportional to the square of their atom/more fraction or also inversely proportional to their partial molar enthalpies of mixing. In fact it is a test, **if experimental data fit-in into a pattern where one of the above relationships are obeyed, the solution is a regular solution**. For a regular solution therefore, the free energy is as for a real solution in terms of activities of components involved but enthalpy and entropy are different as shown in the equations above.

■ **Example 10.1** An alloy of Sn-Ag-Au was made by addition of Au and Ag of total 5 grams at 0°C to a 95 gram molten Sn at 250°C. In one case an alloy Au-Ag (70% Au and 30% Ag) was added whereas in the other a mechanical mixture of Au + Ag was added. The difference in temperature drop of tin in these two cases was 0.5°C. Find out the enthalpy of formation of Au-Ag alloy at 70 : 30 composition. And to what temperature this value of enthalpy of mixing of Au-Ag alloy refer to? Is it 0°, 250° or any other temperature?

Given is sp. heat of Sn-Au-Ag alloy = 0.03 cal/0°/g alloy.

Solution : The process of formation of the final alloy can be imagined to occur as

Process I. Alloy (Au-Ag) $\rightarrow$ + Sn $\rightarrow$ alloy (Sn-Au-Ag)

Process II. Mixture (Au + Ag) $\rightarrow$ + Sn $\rightarrow$ alloy (Sn-Au-Ag)

The difference in these two processes is only as if the alloy Au-Ag is deformed into individual elements Au and Ag then added.

The ΔH^M of Au-Ag (70 : 30) shall thus be additionally required when alloy is added to make up the final required Sn-Au-Ag alloy product. Conversely it can also be imagined that the alloy Au-Ag is formed first from the mechanical mixture and then added to make both the process the same. In that case the ΔH^M of Au-Ag alloy shall be liberated and shall be available for heating the final alloy. In other words the final temperatures reached are not the same. If the final temperature reached is say. T°C in one process,which has to be below 250°C, then in the other process it will be $T \mp 0.5$°C. The thermodynamic process of final alloy formation can be imagined to occur at 0°C in either case. Hence

$$\Delta H_{(\text{process I})} - \Delta H_{(\text{process II})} = \Delta H^M_{(\text{alloy Au-Ag})}$$

Hence,
$$\int_0^T C_P \, dT_{(\text{process I})} - \int_0^{T+0.5} C_P \, dT_{(\text{process II})} = \Delta H^M_{(\text{alloy Au-Ag})}$$

$$\int_0^{T+0.5} C_P \, dT_{(\text{I})} - \int_0^{T} C_P \, dT_{(\text{II})} = \Delta H^M_{(\text{alloy Au-Ag})}$$

In either case the net change shall be

$$\Delta H^M_{(\text{alloy Au-Ag})} = -C \times 0.5 = -0.5 \times 0.03$$

$$= -0.015 \text{ cal/g wt. of alloy Sn-Au-Ag}$$

$$= 0.015 \times 100$$

$$= -1.5 \text{ cal/100 g wt. of alloy Sn-Au-Ag}$$

This is because the total enthalpy change has occurred in 100 grams of the final alloy. This is also equal to the enthalpy change that has occurred in 5 grams of Au-Ag alloy formation in the sum total process. The 5 g Au-Ag alloy has 3.5 g of Au and 1.5 g of Ag or it is equivalent to $\frac{3.5}{197} + \frac{1.5}{108}$ g atoms of Au and Ag respectively.

The total g atoms in the Au-Ag alloy = 0.0177 + 0.0138

or = 0.0315 g atom alloy

$$\Delta H^M = \frac{1.5}{0.0315} = -4760 \text{ cal/g mole alloy Au-Ag}$$

This value of enthalpy of mixing refers to 0°C only because the process has been carried out in such a way that its value at 0°C has come out as an answer. It can not be at any other temperature in this instant case.

10.6 EXCESS FUNCTIONS

The excess thermodynamic function has been defined as the difference between the property of real solution and that if it had behaved as an ideal solution. Thus the general extensive property Z^{xs} (V, H, U, S, A, G, etc.).

$$Z^{xs} = Z_{(\text{real solution})} - Z_{(\text{ideal solution})} \quad \ldots(10.41)$$

If free energy is taken as an illustration then

$$G^{xs} = \Delta G^M_{(\text{real solution})} - \Delta G^M_{(\text{ideal solution})} \quad \ldots(10.42)$$

Putting their values as calculated earlier gives

$$G^{xs} = RT(N_1 \ln a_1 + N_2 \ln a_2 + ... + N_i \ln a_i) - RT(N_1 \ln N_1 + N_2 \ln N_2 + ... + N_i \ln N_i)$$

$$= RT(N_1 \ln \gamma_1 + N_2 \ln \gamma_2 + ... + N_i \ln \gamma_i) \quad ...(10.43)$$

$$= RT \sum_1^i N_i \ln \gamma_i \quad ...(10.44)$$

The excess entropy can be obtained by differentiating the excess free energy that is the above equation with respect to temperature as $\partial G^{xs}/\partial T$ and from this the excess enthalpy H^{xs} can be evaluated as before.

The excess property can be imagined in the form of volume from Fig. 4.1 in Chapter 4 as the difference between the dotted line and the solid line in 4.1(d). It may be +ve or –ve as the case may be. This excess consists of contribution from each individual component of the solution as shown in Fig. 4.1(c). This is known as excess partial molar component and is denoted by $\overline{Z}_i^{xs}$ and can be written for free energy as

$$\overline{G}_i^{xs} = RT \ln \gamma_i \quad ...(10.45)$$

For a solution therefore

$$G^{xs} = N_1\overline{G}_1^{xs} + N_2\overline{G}_2^{xs} + N_3\overline{G}_3^{xs} + ... + N_i\overline{G}_i^{xs} \quad ...(10.46)$$

The excess partial quantities are also interrelated as

$$\overline{G}_i^{xs} = \overline{H}_i^{xs} - T\,\overline{S}_i^{xs} \quad ...(10.47)$$

For a regular solution since

$$\Delta H^M_{(reg)} = RT(N_1 \ln \gamma_1 + N_2 \ln \gamma_2 + ... + N_i \ln \gamma_i)$$

as shown in Eq. (10.35) and therefore,

$$G^{xs} = \Delta H^M_{(reg)}$$

or

$$\overline{G}_i^{xs} = \Delta \overline{H}^M_{i(reg)} = RT \ln \gamma_i \quad ...(10.48)$$

■ **Example 10.2** Calculation of various properties of solution :

If activities are known then various mixing functions can be calculated at any temperature as follows :

Given : The activities of Zn and Cd in an alloy at $N_{Zn} = 0.4$ are 0.3 and 0.65 respectively at 500°K.

Solution : If $N_{Zn} = 0.4$ then $N_{Cd} = 0.6$ in a binary solution.

Hence,

$$\Delta G^M_{500°} = RT(N_{Zn} \ln a_{Zn} + N_{Cd} \ln a_{Cd})$$

$$= 1.987 \times 500 \times 2.303\ (0.4 \log 0.3 + 0.65 \log \text{are } 0.6)$$

$$= -782 \text{ cal/g mole}$$

If the same solution was ideal then

$$\Delta G^{M}_{(real)} = RT\,(N_{Zn} \ln N_{Zn} + N_{Cd} \ln N_{Cd})$$

$$= -\,668 \text{ cal/g mole}$$

hence the excess can be calculated as

$$\Delta G^{xs} = \Delta G^{M} - \Delta G^{M}$$

$$= -\,782 - (-\,668)$$

$$= -114 \text{ cal/g mole}$$

The partial properties can be determined as

$$\Delta \overline{G}_{Zn} = RT \ln a_{Zn} = 1.987 \times 500 \times 2.303 \times \log 0.3$$

$$= -\,1196 \text{ cal/g mole}$$

and
$$\Delta \overline{G}_{Cd} = RT \ln a_{Cd} = 1.987 \times 500 \times 2.303 \times \log 0.6$$

$$= -\,428 \text{ cal/g mole}$$

Now to evaluate partial excess property of each component

$$\Delta \overline{G}_{Zn\,(ideal)} = RT \ln N_{Zn} = 0.4 = -\,910 \text{ cal}$$

$$\Delta \overline{G}_{Cd\,(ideal)} = -\,507 \text{ cal}$$

$$\Delta \overline{G}^{xs}_{Zn} = \Delta \overline{G}_{Zn} - \Delta \overline{G}_{Zn\,(ideal)}$$

$$= -\,1196 + 910 = -\,286 \text{ cal}$$

$$\Delta \overline{G}^{xs}_{Cd} = +\,79 \text{ cal}$$

Now, $\gamma_{Zn} = 0.4/0.3 = 1.33$ and $\gamma_{Cd} = 0.65/0.60 = 1.08$

Hence,
$$\Delta S^{M}_{(reg)} = -\,R(N_{Zn} \ln N_{Zn} + N_{Cd} \ln N_{Cd}) = \Delta S^{M}_{(ideal)}$$

$$= -\,1.987 \times 2.303 \times (0.4 \log 0.4 + 0.6 \log 0.6)$$

$$= -\,1.33 \text{ cal/g mole/0°}$$

Now,
$$\Delta \overline{S}_{Zn} = -\,RT \ln N_{Zn} = 1.987 \times 2.303 \times \log 0.4$$

$$= -\,1.82 \text{ cal/g mole/0°}$$

similarly,
$$\Delta \overline{S}_{Cd} = -\,1.101 \text{ cal/g mole/0°}$$

Then $\Delta H^{M}_{(reg)}$ can be calculated as

$$\Delta H^{M}_{(reg)} = \Delta G^{M}_{(reg)} + T\,\Delta S^{M}_{(reg)}$$

$$= -\,782 + 500\,(-\,1.33)$$

$$= -\,1447 \text{ cal/g mole}$$

It would thus be very clear that how the mixing functions can be evaluated knowing only one parameter like activity. Conversely, if mixing of excess properties are known then the activities or the activity coefficients can be calculated.

10.7 CONCLUDING REMARKS

The free energy, enthalpy and entropy of formation of a real solution are respectively given by Eqs. (10.18), (10.29) and (10.22). Similarly, these for ideal solution are given by, in the same order by Eqs. (10.27), (10.23) and (10.25). The same for regular solution are given, in the same order, by Eqs. (10.18), (10.35) and (10.31) or (10.25).

The partial contribution of free energy of formation of ideal solution is given by Eq. (10.28) and of entropy by Eq. (10.26), the partial enthalpy contribution for ideal solution being zero. The free energy of formation of regular solution is given by Eq. (10.18). Similarly, for regular solution enthalpy is given by Eq. (10.35) and entropy by Eq. (10.31), *i.e.*, Eq. (10.25). The partial contribution in regular solution is given as of free energy by Eq. (10.33), enthalpy of Eq. (10.36) and entropy by Eq. (10.32).

The excess free energy of formation of solution is given by Eq. (10.43) and partial part by Eq. (10.45).

The energy changes on solution formation are by no means insignificant and these arise from contributions of individual components because of the energy changes taking in them individually. Such partial properties and excess properties, when the behaviour differs from the ideal one, do contribute to our understanding of the thermodynamics of solutions.

□□□

Chapter 11

Solutions - III

Applications of Gibb's-Duhem Equation

11.1 INTRODUCTION

The concept of activity of a component in a solution as influenced by the presence of other constituents has been discussed in the previous chapter. It has also been seen that although the practical solution may be multi-component, their properties are mainly estimated from the knowledge of corresponding binary and ternary solutions. It is easier to study and carry out measurements of thermodynamic parameters on binary and ternary solutions. The most commonly measured property of a solution is the activity of their components. Several techniques are available like chemical measurements, emf measurements, vapour pressure measurements and so on, to measure activity of a component in a solution. In a binary solution if activity of a component is measured then mathematically that of the other component can be computed without carrying out the actual measurements. Even if actual measurements are carried out even for the second component mathematical evaluation can be used as confirmatory test on the actual measurements. If by measuring activity of one and determining that of the other by mathematical computation is possible the process can be repeated with more complex mathematics to compute the activity of the third knowing those of the two in a ternary solution. The process can continue for higher order solutions. The accuracy of calculations however does increase with this progress.

The activity is directly related to energy as $d\overline{G}_i = RT \ln a_i$ as per Eq. (4.69). Therefore, even energy properties can be found out for second component when that is known for the first.

The methods of these calculations are described in this chapter.

11.2 DETERMINATION OF PARTIAL MOLAR PROPERTIES

The Gibb's Duhem equation for an extensive property of a binary solution of components 1 and 2 is given by

$$N_1 d\overline{Z}_1 + N_2 d\overline{Z}_2 = 0$$

where, *Z* is any extensive property. It means that

$$d\overline{Z}_1 = -\frac{N_2}{N_1} d\overline{Z}_2 \qquad \text{...(11.1)}$$

This can be integrated provided the variation of $d\overline{Z}_2$ with composition is known for component 2. In a binary when concentration of one component is known $1 - N_1$ shall be the concentration of the other component N_2. The integration can be carried out over the limits to be fixed. The natural choice is of lower limit as that when $N_2 = 0$, *i.e.*, $N_1 = 1$ or $d\overline{Z}_1 = 0$ and the upper limit is that when $N_1 = N_1$ at which the $d\overline{Z}_1$ is to be found out. The integration of Eq. (11.1), thus gives

$$\int_{N_1=1}^{N_1=N_1} \times d\overline{Z}_1 = -\int_{N_1=1}^{N_1=N_1} \times d\overline{Z}_2 \qquad \text{...(11.2)}$$

or

$$\overline{Z}_{1(N_1=N_1)} - \overline{Z}_{1(N_1=1)} = \int_{N_1=1}^{N_1=N_1} \times \frac{N_2}{N_1} d\overline{Z}_2$$

where, $N_1 = 1$, $\overline{Z} = 0$ and therefore,

$$\overline{Z}_{1(N_1=N_1)} = -\int_{N_1=1}^{N_1=N_1} \times \frac{N_2}{N_1} d\overline{Z}_2 \qquad \text{...(11.3)}$$

This integral can be solved graphically to compute the value of $\overline{Z}_1$ at $N_1 = N_1$ concentration. This is illustrated before with the help of an example and the calculations in involved therein.

■ **Example 11.1** Let us assume that the following data were obtained for partial molar energy of mixing of Ag-Si alloys at 1150°C.

Solution :

N_{Si}	1.0	0.9	0.8	0.7	0.6	0.5	0.4	0.3	0.2	0.1	00	
$\overline{G}_{Si}$	0.0	90	120	280	400	550	800	1150	1400	1800	0	cal/mole

Let us say that partial molar free energy of Ag at 0.2 atom fraction is required to be computed. In the equation (11.2) or (11.3) component 1 is Ag and 2 is Si. It therefore, reduces to :

$$\int_{N_{Ag}=1}^{N_{Ag}=0.2} \times d\overline{G}_{Ag} = \int_{N_{Ag}=1}^{N_{Ag}=0.2} \times \frac{N_{Si}}{N_{Ag}} d\overline{G}_{Si}$$

or this will reduce down to

$$d\overline{G}_{Ag=0.2} = -\int_{N_{Ag}=1}^{N_{Ag}=0.2} \times \frac{N_{Si}}{N_{Ag}} d\overline{G}_{Si}$$

The available data will have to be processed further as

N_{Si}	1.0	0.9	0.8	0.7	0.6	0.5	0.4	0.3	0.2	0.1	00	
$\overline{G}_{Si}$	0.0	0.1	0.2	0.3	0.4	0.5	0.6	0.7	0.8	0.9	00	
$\frac{N_{Si}}{N_{Ag}}$		9.0	4.0	2.33	1.50	1.0	0.66	0.43	0.25	0.11		
$\overline{G}_{Si}$	0	90	120	280	400	550	800	1150	1400	1800	0	cal/mole

A graph is plotted with $\overline{G}_{Si}$ on the x-axis and the ratio $\frac{N_{Si}}{N_{Ag}}$ on the y-axis as shown in Fig. 11.1.

The area on the graph under the curve and bound by the x-axis between the limits $N_{Ag} = 1$ to $N_{Ag} = 0.02$ is shown by thatched marks. The area can be worked out graphically but since it extends in the form of a tail to $-\infty$, *i.e.*, asymptotic form, it can not be determined accurately. In the present

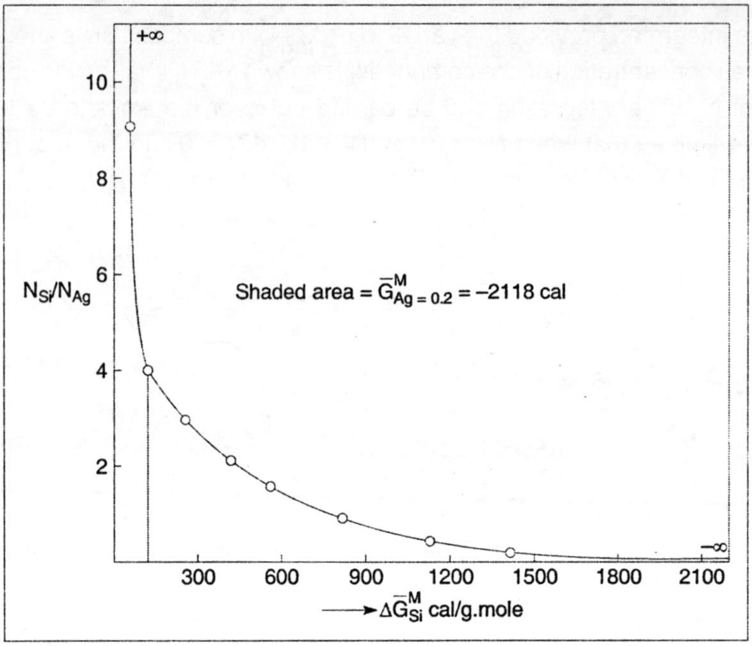

Fig. 11.1 Plot to evaluate the Gibb's-Duhem integral for evaluation of partial molar property.

case it comes out to be 2118 cal/mole and hence the partial molar free energy of mixing of Ag at $N_{Ag} = 0.2$ is −2118 cal/mole.

The values of $\overline{G}^M_{Ag}$ at various concentrations can be computed likewise for each composition separately, **by computing each time the area as above.**

11.3 DETERMINATION OF ACTIVITIES

In the case of binary solutions it is experimentally readily possible to determine activity of a component, particularly the solute, in a solution using a varieties of techniques but it is rather difficult to determine experimentally the activity of the solvent component at the same composition. This difficulty is more pronounced in the case of dilute solutions. The integration of Gibb's Duhem equation comes out handy to determine the activity of the other component, usually the solvent, when that of the solute component is known by actual measurements. It has already been seen that activity is nothing but a function of partial molar free energy and hence the method of integration of Gibb's Duhem equation inthe previous section should be applicable for estimation of activity as well. In this case the Gibb's Duhem equation is to be transformed into an equation correlating activities of the two components in a binary solution. By taking partial molar free energy as the extensive property the Gibb's Duhem equation can be written as :

$$N_1\, d\overline{G}_1 + N_2\, d\overline{G}_2 = 0$$

Putting $d\overline{G}_1 = RT \ln a_i$ and dividing by RT

gives
$$N_1\, d \ln a_1 + N_2\, d \ln a_2 = 0 \qquad \text{...(11.4)}$$

eliminating natural logarithm

$$N_1\, d \log a_1 + N_2\, d \log a_2 = 0 \qquad \text{...(11.5)}$$

or
$$d \log a_1 = -\frac{N_2}{N_1} d \log a_2 \quad ...(11.6)$$

Integrating again between the limits $N_1 = 1$ at the lower limit to $N_1 = N_1$ as the upper limit gives
$$\int_{N_1=1}^{N_1=N_1} d \log a_1 = -\int_{N_1=1}^{N_1=N_1} \frac{N_2}{N_1} d \log a_2 \quad ...(11.7)$$

or
$$(d \log a_1)_{N_1=N_1} - (d \log a_1)_{N_1=1} = -\int_{N_1=1}^{N_1=N_1} \frac{N_2}{N_1} d \log a_2$$

since when $N_1 = 1$, $a_1 = 1$ and hence $\log a_1 = 0$ thus
$$(d \log a_1) = -\int_{N_1=1}^{N_1=N_1} \frac{N_2}{N_1} d \log a_2 \quad ...(11.8)$$

A graph plotted between N_2/N_1 and log a and the area under the curve bound by the x-axis over the range $N_1 = 1$ to $N_1 = N_1$ is estimated to obtain the value of $\log a_1$ $N_1 = n_1$. The integration is carried out between limits upto different N_1 concentrations. The curve approaches at one end + infinity when $N_1 = 0$ to – infinity when $N_2 = 0$ at the other end. The difficulties encountered will be clear by taking a numerical example and solving it as follows :

■ **Example 11.2** Let us take the measurements of activities of Ni in Fe-Ni alloys at 1600°C made by G.R. Belton and R.J. Fruehan (J. Phys. Chem. 1967), which are shown in horizontal columns 1 and 2. Let us try to calculate the activity of iron at $N_{Fe} = 0.9$.

Solution :

N_{Ni}	1.0	0.9	0.8	0.7	0.6	0.5	0.4	0.3	0.2	0.1	00
a_{Ni}	1.0	0.89	0.766	0.62	0.485	0.374	0.283	0.207	0.137	0.067	—

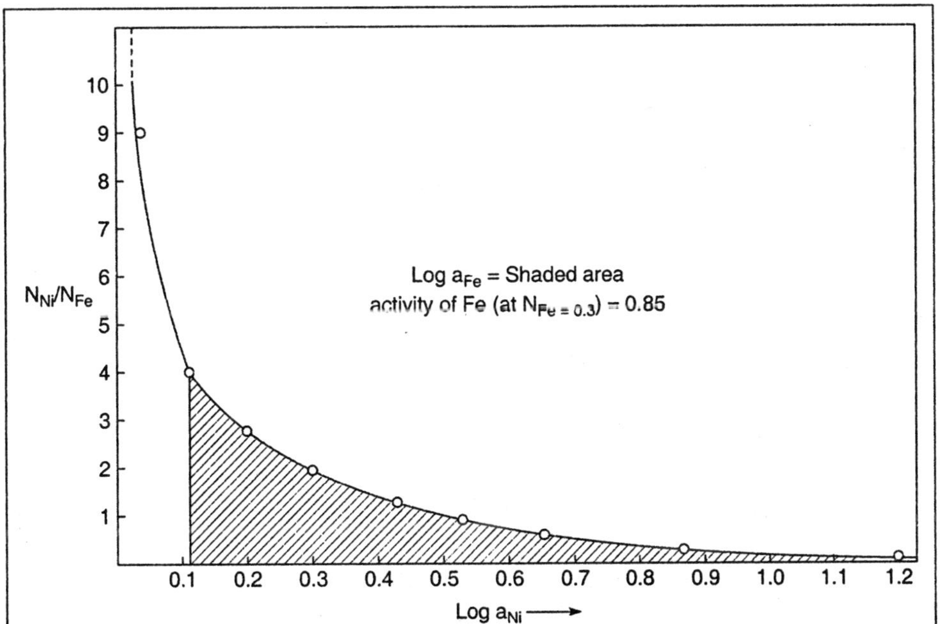

Fig. 11.2 Graphical determination of activity using the Gibb's-Duhem integration.

Now, calculated values with the same columns continued

N_{Fe}	1.0	0.9	0.8	0.7	0.6	0.5	0.4	0.3	0.2	0.1	00
$\frac{N_{Ni}}{N_{Fe}}$	—	9.0	4.0	2.33	1.50	1.0	0.66	0.43	0.25	0.11	—
$\log a_{Ni}$	–	0.0506	0.1157	0.2076	0.3142	0.4271	0.5482	0.6840	0.8664	1.1739	

In order to solve the integral the data are processed to obtain horizontal columns 3, 4 and 5 and graph is plotted with $\log a_{Ni}$ on x-axis and N_{Ni}/N_{Fe} on the y-axis as shown in Fig. 11.2.

The area under the curve bound by x-axis between the limits $N_{Ni} = 1$ to $N_{Ni} = 0.1$ is estimated as the value of the integral in Eq. (11.8). It is observed that the curve extends to – infinity when $N_{Ni} = 1$ and hence required area has to be calculated strictly upto – infinity limit. This is practically just not possible and hence the calculations are in error. However, with all these errors the value of $\log a_{Fe}$ at $N_{Fe} = 0.2$ is equal to the area shown by thatchings and is equal to 0.85.

11.4 ESTIMATION OF ACTIVITY COEFFICIENT

If activity can be interrelated through Gibb's - Duhem equation, activity coefficient must also be similarly interrelated. As derived in the previous section in Eq. (11.4)

$$N_1 \, d \ln a_1 + N_2 \, d \ln a_2 = 0$$

Since in a binary $\quad N_1 + N_2 = 1$

hence $\quad dN_1 + dN_2 = 0$

altering the forms suitably $\quad N_1 \dfrac{dN_1}{N_1} + N_2 \dfrac{dN_2}{N_2} = 0$

it also means $\quad N_1 \, d \ln N_1 + N_2 \, d \ln N_2 = 0$

from definition $a_i = N_i \gamma_i$, *i.e.*, $\quad \log a_i = \log N_i + \log \gamma_i$

or $\quad \log a_i - \log N_i = \log \gamma_i \qquad ...(11.9)$

Subtracting Eq. (11.9) from Eq. (11.4) leads to

$$N_1 \, d \log \gamma_1 + N_2 \, d \log \gamma_2 = 0 \qquad ...(11.10)$$

$$d \log \gamma_1 = -\frac{N_2}{N_1} d \log \gamma_2 \qquad ...(11.11)$$

As before integrating between the limits

$$N_1 = 1 \quad \text{to} \quad N_1 = N_1$$

$$\int_{N_1 = 1}^{N_1 = N_1} d \log \gamma_1 = -\int_{N_1 = 1}^{N_1 = N_1} \frac{N_2}{N_1} d \log \gamma_2 \qquad ...(11.12)$$

$$(\log \gamma_1)_{N_1 = N_1} - (\log \gamma_1) = -\int_{N_1 = 1}^{N_1 = N_1} \frac{N_2}{N_1} d \log \gamma_2$$

Since at $N_1 = 1$, $a_1 = 1$ and hence $\gamma_1 = 1$ and therefore $\log \gamma_1 = 0$

therefore, $\quad (\log \gamma_1)_{N_1 = N_1} = -\displaystyle\int_{N_1 = 1}^{N_1 = N_1} \frac{N_2}{N_1} d \log \gamma_2 \qquad ...(11.13)$

This can be solved graphically by plotting N_2/N_1 vs. $\log\gamma_2$ and the area under the curve bound by the x-axis between the limits $N_1 = 1$ to $N_1 = N_1$ shall give the value of $\log\gamma_1$ and from which γ_1 can be calculated at $N_1 = N_1$. Such values at any other composition can be calculated likewise. Knowing γ_1, the activity values can be calculated through the relationship $a_i = N_i \cdot \gamma_i$. The activity data of Ni in Fe-Ni system at 1600°C shown in the previous section can be processed further with respect to Eq. (11.13) to understand the method of this calculation more precisely.

N_{Ni}	1.0	0.9	0.8	0.7	0.6	0.5	0.4	0.3	0.2	0.1	00
a_{Ni}	1.0	0.89	0.766	0.62	0.485	0.374	0.283	0.207	0.137	0.067	—
γ_{Ni}	-	0.988	0.95	0.885	0.808	0.748	0.707	0.69	0.68	0.67	
$\log\gamma_{Ni}$	-	−0.0052	−0.022	−0.053	−0.09	−0.126	−0.150	−0.161	−0.167	−0.173	
N_{Fe}	0.0	0.1	0.2	0.3	0.4	0.5	0.6	0.7	0.8	0.9	00
$\frac{N_{Ni}}{N_{Fe}}$		9.0	4.0	2.33	1.50	1.0	0.66	0.43	0.25	0.11	—

The graph is plotted with these data with $\log\gamma_{Ni}$ on x-axis and N_{Ni}/N_{Fe} on the y-axis. The graph is shown Fig. 11.3. The activity coefficient of iron at 0.3 atom fraction composition is

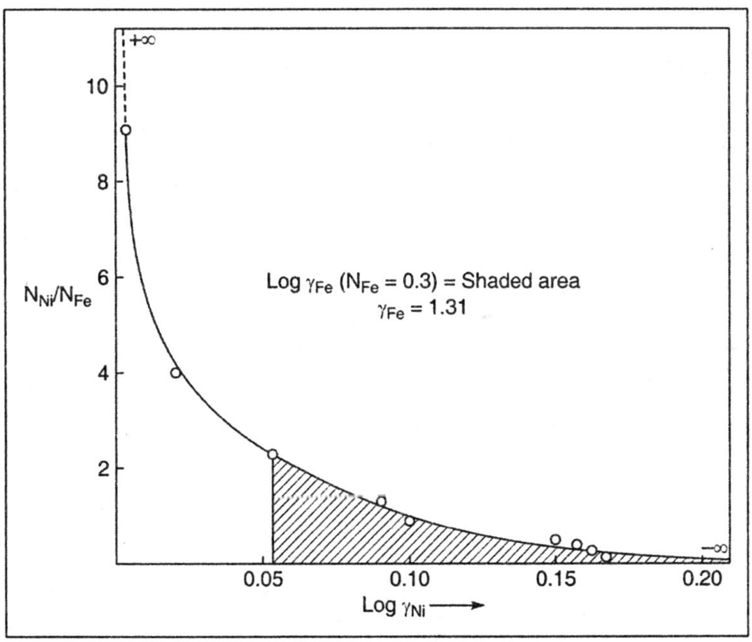

Fig. 11.3 Graphical determination of activity coefficient using the Gibb's-Duhem integration.

equivalent to the thatched area on the graph and which is the value of the integral as desired by Eq. (11.13). The area is equivalent to $\log\gamma_{Fe} = 0.01186$ and hence $\gamma_{Fe} = 1.31$.

11.5 ESTIMATION OF ACTIVITY COEFFICIENT USING ALPHA FUNCTION

From the previous section

$$N_1 \, d \log \gamma_1 + N_2 \, d \log \gamma_2 = 0 \quad \text{as in Eq. (11.10)}$$

dividing both by dN_2

$$N_1 \frac{d \log \gamma_1}{dN_2} + N_2 \frac{d \log \gamma_2}{dN_2} = 0 \quad ...(11.13)$$

or

$$d \log \gamma_1 = -\frac{N_2}{N_1} \cdot \frac{d \log \gamma_2}{dN_2} dN_2 \quad ...(11.14)$$

As before this can be integrated between the limits $N_1 = 1$to $N_1 = N_1$ gives

$$\int_{N_1=1}^{N_1=N_1} \times d \log \gamma_1 = -\int_{N_1=1}^{N_1=N_1} \times \frac{N_2}{N_1} \frac{d \log \gamma_2}{dN_2} dN_2 \quad ...(11.15)$$

By definition the term

$$\frac{\ln N_i}{(1-N_1)^2} \text{ is the } \alpha \text{ (alpha) function for}$$

binary solution and as has been observed

$$\alpha_1 = \frac{\ln \gamma_1}{N_2^2} \quad \text{and} \quad \alpha_2 = \frac{\ln \gamma_2}{N_2^2}$$

and $\alpha = \alpha_1 = \alpha$ and hence

$$\alpha = \frac{\ln \gamma_1}{N_2^2} = \frac{\ln \gamma_2}{N_2^2} \quad ...(11.16)$$

and by processing this equation it can be shown that

$$\log \gamma_1 = -\alpha \, N_1 \, N_2 - \int_{N_1=1}^{N_1=N_1} \alpha \, dN_1 \quad ...(11.17)$$

therefore, if α, *i.e.*, $\log \gamma_2/(1-N_2)$ is plotted vs. N_1 the area under the curve bound by x-axis between the limits $N_1 = 1$ and $N_1 = N_1$ can be obtained and if from this computed value of $-\alpha \, N_1 N_1$ is subtracted, the value of $\log \gamma_1$ is obtained.

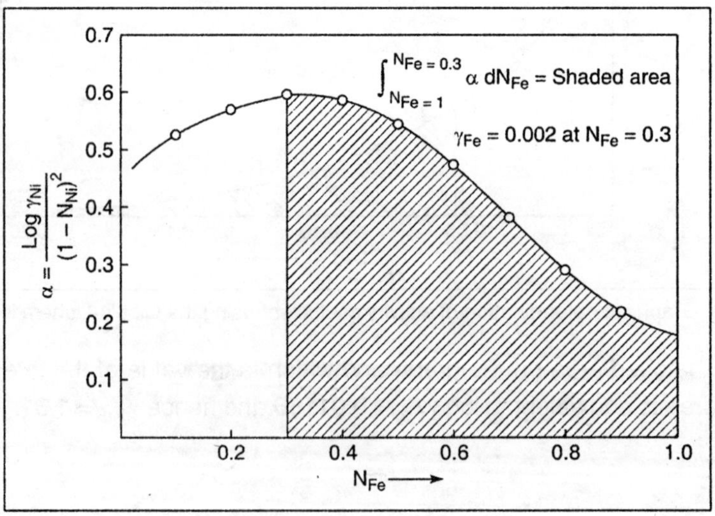

Fig. 11.4 Graphical evaluation of activity coefficient using alpha-function.

There is no asymptotic shape of this curve and exact area can be found out to get the activity coefficient accurately. This is shown in Fig. 11.4 using the data of the previous section for activity of nickel to work out the activity coefficient of iron at $N_{Fe} = 0.3$. This α value at $N_{Fe} = 0.3$ can be evaluated.

The value of the integral can be assessed from the graph in Fig. 11.4 and is shown by shaded area. The above equation (11.17) can now be solved to work out the value of log γ at $N_{Fe} = 0.3$ and from there the value of activity coefficient of Fe works out to be 0.002 at iron atom fraction of 0.3.

Alternatively if activity coefficient of a component in a solution is expressed by an empirical equation then if can be processed mathematically to get the activity coefficient of the second component.

11.6 CONCLUDING REMARKS

The Gibb's - Duhem equation is thus very useful in evaluating activity, its coefficient or any extensive property of a component in solution when that of the other component is known over a range of composition. Although it is tedious but is useful particularly when the measurements for the other component are not accurately measurable by experimentation.

□□□

Chapter 12

Equilibria in Phase Diagrams

12.1 INTRODUCTION

In a single component system, *i.e.*, in pure substances the phases that can exist under appropriate conditions of temperature and pressure are the solid or liquid or gas forms of the pure substance. For a given amount of pure substance if P and T was fixed the volume is automatically fixed. Even if amount of substance is varied T and P are potential terms and hence decide the equilibria independent of mass of the substance. The areas where each one of these phases exist are therefore available from T-P diagrams as shown in Fig. 6.2. The T-P relationships also indicate the conditions when solid-liquid, solid-gas and liquid-gas shall co-exist. This is confirmed from G-T relationship in terms of associated minimum free energy for the stable phase or phases as in Fig. 6.3. Although these relationships do indicate phases in equilibrium in a single component system, these are not called equilibrium diagrams.

Equilibrium or phase diagrams however mean graphical representation of stable equilibrium phase regions that exist with respect to temperature and composition in a system with minimum of two components wherein at least one component is in a condensed form. The pressure is fixed at one atmosphere normally in plotting these diagrams. The equilibrium diagrams are often known in terms of the number of components involved in the system. If only two components are present, it is binary equilibrium or phase diagram, if three are present it is ternary diagram and, if four, it is quaternary diagram and so on. A binary diagram can be depicted readily as a two dimensional diagram with composition on the x-axis and temperature on the y-axis. A ternary can be indicated in a triangular form as a two dimensional representation of the three dimensional situation at fixed temperature. The quaternary form is still complicated. In a more than four, that is truely multi-component system, it is not at all possible to construct he diagram as a variation of temperature and composition and hence for a fixed composition the phases are ascertained or it is reduced to a pseudo-ternary system to be depicted in a triangular form.

The binary diagrams are by far the most popular, useful, readily legible and available forms of phase diagrams. As a preliminary introduction to the subject of energy considerations and phase equilibria, the discussion here is confined to binary phase diagrams only.

A binary system even in a condensed form may exhibit, one or two liquid phases, one or more solid solution phases, one or more intermetalic or chemical compounds, one or more mechanically mixed phases as eutectics, magnetic and non-magnetic phases, and so on. In other words the diagram may depict regions of stabilities of several of such phases. The simplest diagram with one liquid and one solid solution, where complete solid and liquid solubility between the two components exist, shall have the region of stabilities of liquid and solid solution phases and separating these two phase regions, a area where both phases co-exist in in equilibrium. There will have to be a two phase region, separating two single phase regions. When one phase changes to the other on heating or cooling, the gradual changeover through the two phase region is thus necessary and without this direct changeover from one to the other phase is just not possible.

The equilibrium diagram of two completely immiscible components even in liquid state, *e.g.*, Pb-Fe, W-Pb etc. are of no interest because no other phases, except the two pure component phases, mechanically in contact with each other, exist there. Therefore phase diagrams where at least a partial liquid solubility of component, one in the other, is possible, are of relevance. Alternatively at least complete liquid solubility without any solid solubility may also be relevant.

The phase diagrams are constructed using equilibrium measurements and subsequent phase analysis to be carried out by either chemical, X-ray, metallography and such other techniques. Dilatometric or differential gravimetric (D-G) or differential thermogravimetric (D-T-G) techniques are used to know the temperatures at which phase changes occur. Those measurements are carried out over the required temperature and composition ranges to complete the diagram. These measurements are carried out reversibly to ensure equilibrium. The fact that these are equilibrium diagrams means that they indicate the equilibrium phases existing in a given temperature-composition range under isobaric conditions with respect to the composition. The second law clearly establishes that for a system of definite composition at a fixed temperature and pressure, the equilibrium stable phase is the one which possesses lowest free energy. Equally true would be that when two phases coexist in this diagram the chemical potentials, $\overline{G}_i$ for a component in the coexisting phases must be the same. Alternatively the activities of the same component in the co-existing phases must be the same since free energy and activity are directly related to each other. In other words the phase diagrams must show appropriate free energy composition and activity composition relationships in conformity with the laws of thermodynamics.

The phase diagrams are evaluated by assuming cooling of certain composition of the system from where it is fully liquid. As this cools no change will occur till the temperature touches the liquids boundary (above which only liquid exists). Every phase diagram would pose this situation. At the liquidus temperature, since by its definition, that solid phase shall separate out from the liquid. The equilibrium involved is

$$\text{Liquid} \leftrightarrow \text{Solid}$$

and solid phase can be almost the pure component, a solid solution, an intermetallic compound or a mechanical mixture of components known as a eutectic. This will continue till all the liquid phase is transformed into solid. So the free energy of formation of one of these solid phases, when minimum, give the clue regarding the separate phase(s) or equilibrium phase(s) and the temperature range over which it(these) shall be stable.

Similarly the solid phase so formed on further cooling can undergo phase changes as :

1. Solid solution $\leftrightarrow$ two solids
   (eutectoid change)
2. Solid solution $\leftrightarrow$ another solid solution
3. Solid solution $\leftrightarrow$ a liquid and a solid

As the above free energy or activity consideration shall apply to this and such other equilibria, involved in the given component, as it cools to almost room temperature or even below it if required. Any other changes also can be treated likewise.

In order to correlate the phase diagram with free energies of component phases and the relevant activity of components a few simple diagrams are considered below.

12.2 FREE ENERGY OF FORMATION OF SOLUTION

There are systems where complete solubility over entire composition range in liquid shall exist. But many ceramic systems like alkaline earth-metal oxide-silica systems exhibit only partial liquid solubility. There are many where only one solid solution exists over the entire composition ranges like Cu-Ag, Cu-Ni, Pb-Au etc. But many others exhibit only partial solid solubility at each end. In the extreme the solid solubility is so negligibly small that almost pure component separate out without any solid solution formation at each end. The Pb-Sn is a typical system of this type wherein the liquid freezes into a fine mixture of Pb and Sn as eutectic or a certain fixed composition and the excess of any component as a separate phase of that pure component.

The evaluation of a solution phase thus start with determination of free energy of formation of a solution, initially in liquid state and then in solid solution form as well. The ΔG^M, ΔH^M and ΔS^M for the formation of a solution from components 1 and 2 have already been evaluated in Chapter 10 and can be reproduced as

$$\Delta G_{ideal} = -T\,\Delta S_{ideal}$$
$$= -RT\,[N_1 \ln N_1 + N_2 \ln N_2]$$

since $\Delta G_{ideal} = 0$ for ideal solution. However, if solution is not ideal, *i.e.*, $a_i \neq N_i$ but $a_i = \gamma_i\, N_i$ or $h_i = f_i$ (wt %) and hence

$$\Delta G^M_{real} = \Delta H_{real} - T\,\Delta S_{real}$$

The actual value of ΔG^M depends on whether ΔH^M is –ve or +ve, *i.e.*, exothermic or endothermic. But when solution forms ΔH^M will have to be –ve, being a natural process. Since the interaction amongst the components manifest as –ve of +ve values of ΔH^M,the activities of components shall show corresponding –ve or +ve deviation from ideality. If all the information is known the ΔG^M can be evaluated for any composition of solution at any temperature. ΔG^M is therefore a function of temperature and composition (pressure 1 atmos. fixed).

The free energy of formation of a solution is shown in Fig. 12.1 under varieties of conditions. The curve at 'A' is ideal solution formation and 'B' is for the –ve and 'C' is for the +ve deviation from ideality. Depending upon the actual value of ΔH^M, which will definitely be related to interaction of components the real ΔG^M shall acquire the value as a resultant of ΔH^M and $T\,\Delta S^M$ factors.

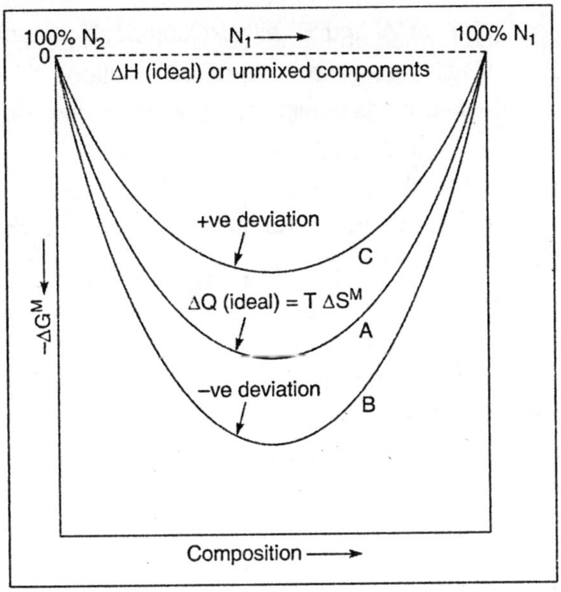

Fig. 12.1 Free energy of formation of an ideal solution as a function of composition along-with those of real solutions to indicate the positive and negative deviation from ideality.

The fact that components from solution, being a natural process, is itself an indication of –ve value of ΔG^M for the solution. ΔH^M also will have to be –ve as the natural process is exothermic in nature. The ΔS^M will however be a +ve quantity for the same reason. Another interesting part is that the ΔG_i^M can be evaluated by drawing a tangent at the corresponding point of composition on the ΔG^M curve, and the intercepts of this tangent as shown in Fig. 12.1 on axis at $N_1 = 0$ and $N_2 = 0$ give the values of $\Delta\overline{G}_1^M$ and $\Delta\overline{G}_1^M$. At $N_1 = 0$ means $N_2 = 1$ and at $N_2 = 0$ means $N_1 = 1$, the slope will have to be minus infinity and hence the curve at these two extremities must be vertical since at $N_1 = 1$, $\Delta\overline{G}_1^M = 0$ and $\Delta\overline{G}_2^M = -\infty$ and at $N_2 = 1$, $\Delta\overline{G}_2^M = 0$ and $\Delta\overline{G}_1^M = -\infty$.

The –ve ΔG^M value for any such solution is a sure indication of the stability of liquid solution and upto whatever extent it is so, the solubility limit shall extend. The Fig. 12.1 shows the state of affairs when liquid solubility is exhibited over entire range of composition. No other form of phase, like say a mixture of components 1 and 2 shall be stable. Even if the ΔG^M on this mixture is evaluated its values would be less –ve than the corresponding value ΔG^M for solution as shown in Fig. 12.1.

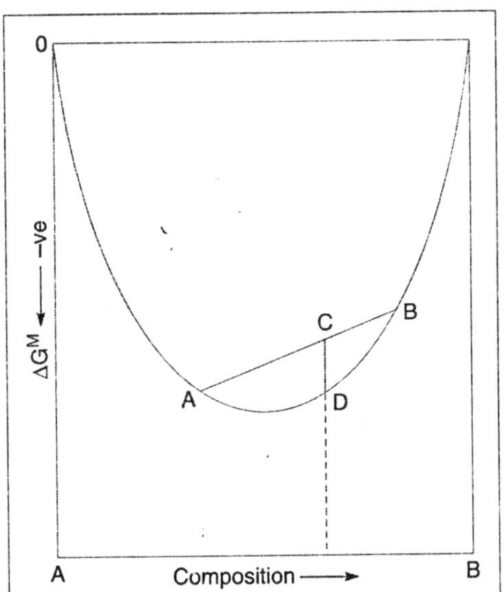

Fig. 12.2 Shows free energy of mixing as it varies with composition for a solution.

This can be proved from Fig. 12.2 which is reproduction of the Fig. 12.1. For ΔG^M as a generalised case, not specifically ideal or otherwise. Let us imagine a mixture of two solutions of composition 'A' and 'B' with the composition average being formed at 'C'. The ΔG^M will be the total of ΔG of these two component solutions as given by point A and B respectively. The sum total value shall be as depicted by C which is much lower than (less –ve) point 'D' which gives ΔG^M of solution of A and B solutions. Hence solution D is more stable than mixture represented by C. This holds true for entire range and therefore solution phase shall be more stabler than mere mixture.

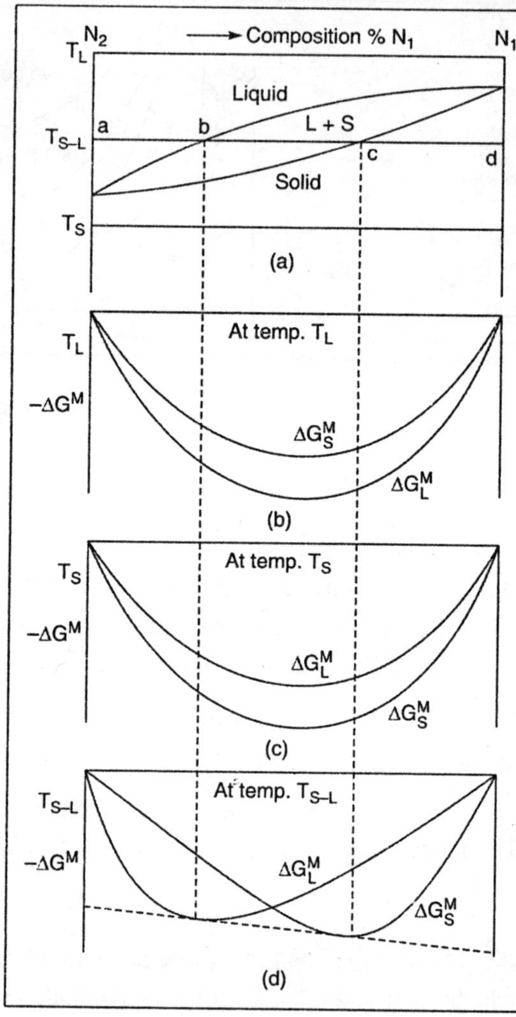

Fig. 12.3 Free energy-composition relationships for various phases at different temperatures for binary systems exhibiting complete solid solubility.

Fig. 12.3 gives the free energy of formation of solid as well as liquid solution over the entire composition range at three different temperatures for a system which exhibits solubility in both liquid and solid state over the entire composition range. The ΔG^M for solid solution and liquid solution at temperature T_S and T_L make interesting reading. At temperature T_L only liquid solution can exist in the phase diagram. The values of ΔG_L^M and ΔG_S^M have been plotted in Fig. 12.3(b). The ΔG_L^M has all through values more negative than those of ΔG_S^M values for any composition, meaning thereby that liquid solution is stabler than solid solution. The phase diagram conforms this. Similarly the argument for temperature T_S shall indicate solid solution to be the stable phase as shown in (c). At temperature T_{S-L} the temperature axis cuts the liquids at 'b' and solidus at 'c' as shown in (a). The ΔG^M vs. composition in the diagram indicates that for composition left of b, liquid is stable and for composition right of c, solid solution is stable. In between b and c, both ΔG_S^M and ΔG_L^M are the same and hence they co-eixst. The phases in the diagram are thus confirmed on the basis of free energy considerations.

12.3 FREE ENERGIES OF SYSTEMS WITH ONLY PARTIAL SOLID SOLUBILITY

The system is typically shown in Fig. 12.4. The ΔG^M of solid solution for such a system is like that in Fig. 12.1. Let us work out the free energies of formation of solid solution and liquid solution at various temperatures, as depicted by the horizontal dotted line for the corresponding values. The ΔG^M for liquid, solid solution α and solid solution β have been plotted at temperature T_{S-L} in Fig. 12.4(b). It is apparent that between the compositions 'b' and 'c' the ΔG^M liquid phase has the lowest value as compared to ΔG^M for α and β solid solutions and hence liquid phase is stable in this region as been in the phase diagram. Also the ΔG^M_α has lower value on the left of composition b as it is this region where α separates out from the liquid solution. Similarly, ΔG^M_β shall indicate the precipitation of β solid solution beyond right of composition c.

It is equally clear that at T_{S-L} in between b and c that ΔG^M for $\alpha - \beta$ solid solution is more than that for liquid solution and hence α and β, individually are not stable phases in this region. As the temperature is lowered down below T_{S-L} the miscibility gap decreases and eventually at T_E both α and β have ΔG^M less than that for liquid and hence are stable. But for the region between 'g' and 'f', both ΔG^M_α and ΔG^M_β are same whereas for composition less than point 'g' it is $\Delta G^M_\alpha < \Delta G^M_\beta$ and above point 'f' $\Delta G^M_\beta > \Delta G^M_\alpha$ indicating the α and β stability ranges at either end of the phase diagram. If temperature is further lowered the overlapping of α and β lines extend further thereby decreasing the stability region of α and β as independent phases. This is obvious because of the slope of $\alpha/(\alpha + \beta)$ and $\beta/(\alpha + \beta)$ phases being sloping towards lower solid solubility ranges.

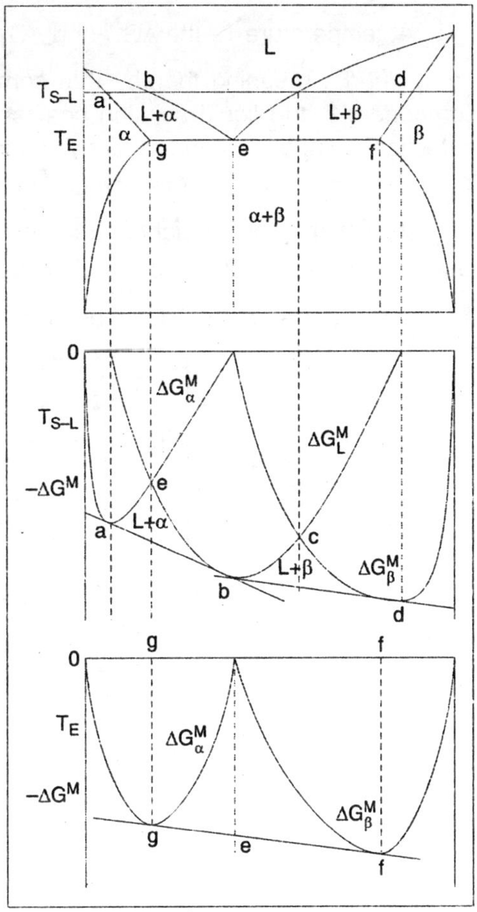

Fig. 12.4 Free energy-composition relationships for various phases in a binary phase diagram with partial solid solubility.

12.4 FREE ENERGY OF PURE EUTECTIC SYSTEM

A typical pure eutectic system is shown in Fig 12.5(a). Let us consider the temperature which is between the m.p. of the two constituents as shown by the line $T_{S-L(1)}$. The corresponding free energy values, ΔG^M_L and ΔG^M_S, are plotted in Fig. 12.5(b). Since temperature is below the m.p. of component 1 and since no solid solution is formed at 'a', solid-1 must precipitate out on further cooling. In other words $\Delta G^M_{S(1)}$ must have a lower value than ΔG^M_L. If a tangent is drawn at corresponding point 'a' on the free energy curve the intercept will give partial, in this case of pure component, free energy which is less than that of solution as $\Delta G_1 = \Delta \overline{G}^M_1$. This is shown in Fig. 12.5(c).

At a lower temperature $T_{S-L(2)}$ which is lower than the lowest m.p. of the two constituent, the situation as shown in Fig. 12.5(b) will qualitatively exists at the other end of the diagram as well.

At temperature T_E the ΔG_L^M and ΔG_1 are overlapping on the left side of eutectic composition, *i.e.*, point 'E', meaning thereby that component 1 and liquid coexist and on the right similarly, component 2 and liquid solution coexist. Interestingly the pure phase 1 is unstable on right of eutectic E composition and similarly phase 2 is unstable on the left of E. If situation at temperature T_E and composition E is considered then ΔG_L^M, and ΔG_1 shall overlap on the left of E and ΔG_L^M, and ΔG_2 shall overlap on the right of E to indicate their coexistence, which is the case. The ΔG_1 will be less –ve on the right of E and therefore, below ΔG_L^M curve and ΔG_2^M will be less –ve on the left of E, and therefore, below ΔG_L^M curve as is so obvious in Fig. 12.5(d).

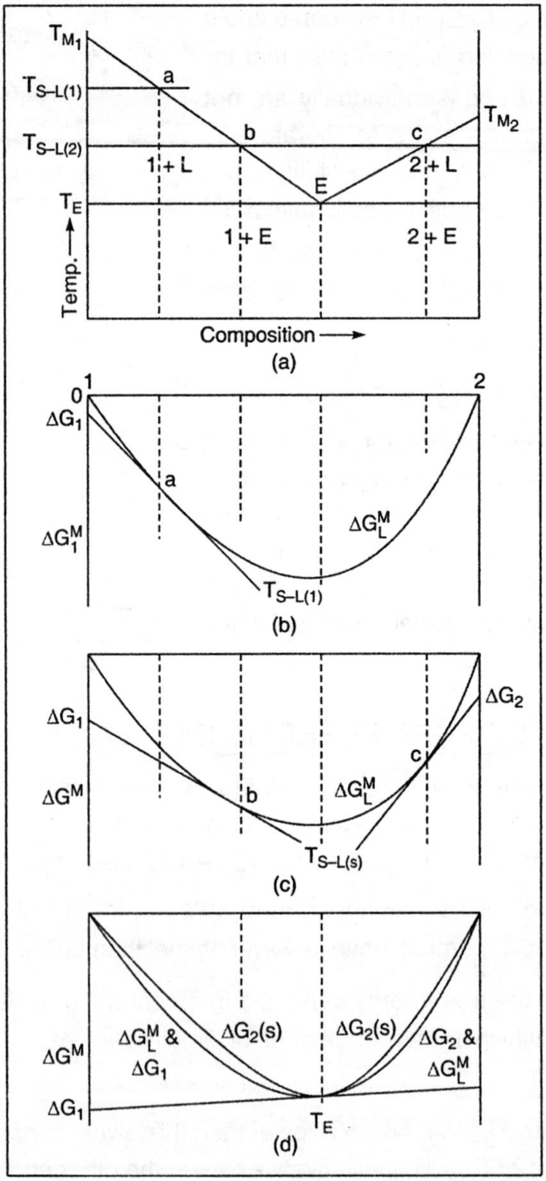

Fig. 12.5 Free energy-composition relationships at various temperatures for a pure eutectic system vis-a-vis the phase diagram:

12.5 LIMITING CONDITIONS FOR MISCIBILITY GAP IN PHASE DIAGRAMS

When solubility exists in liquid state the ΔG_L^M is always –ve, as it is a natural process. But if the constituent components show a +ve deviation from Roultian behaviour then the ΔG_L^M values become less negative and if the +ve deviation is sufficiently large then ΔG_L^M may even have +ve values over certain range of composition. It means there is a tendency not to form solution and exhibit liquid immiscibility in the region where ΔG_L^M is +ve. The normal convex downward peak of ΔG_L^M, when –ve convex upwards as shown in Fig. 12.6.

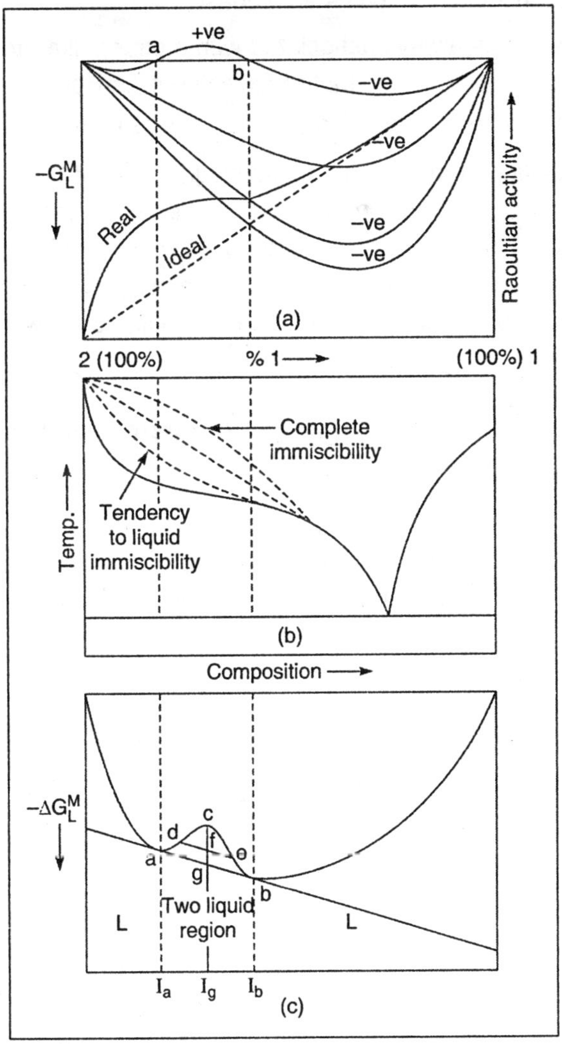

Fig. 12.6 The free energy-composition relationships to explain the conditions when liquid immiscibility becomes imminent.

It is clear that for solutions of compositions 1_a and 1_b the ΔG_L^M are the same since the tangent is same and hence must be coexistent in equilibrium. The ΔG_L^M however increases beyond 1_a with composition and after attaining a peak in convex upward form reduce to attain the maximum value at 'b' as shown in Fig. 12.6(b). For any composition of solution between the points 'a' and 'b' the composition can be imagined as a mixture of two solutions, say for example as of 'd' and 'e' with the effective gross composition shown by 'f'. The ΔG_L^M for composition 'f' is thus that given by point 'c'. But as a mixture its real value is depicted by 'f'. In other words the system has tendency to exist as two liquids rather than a homogeneous solution having ΔG_L^M corresponding to point c because, free energy corresponding to point f is lower than that for 'c'. The stable configuration between composition of 'a' and 'b' are two liquids of composition 'a' and 'b'. On the left of 'a' and right of 'b' homogeneous liquid solutions shall exist as stable phases.

The liquid miscibility gap is always associated with a strong like-atom/molecule attraction or +ve deviation from Raoultian behaviour. Mathematically, it can be deduced that or $d\overline{G}/dN_1$ or d^2G/dN_1^2 or d^3G/dN_1^3 must all be zero for miscibility to occur as composition changes from one end to the other in a binary solution. It means the phase diagrams must have a liquidus, not varying in temperature with composition as shown in Fig. 12.6 between compositions 'a' and 'b'. Also since activity is directly proportional to free energy the activity also should either show to +ve variation with composition or it should not show any variation, *i.e.*, invariance with composition. This is shown in Fig. 12.6(a) for compositions between 'a' and 'b'. That it should no show inverse deviation at some stage as it varies with composition goes without saying.

12.6 FREE ENERGY OF FORMATION OF COMPOUNDS IN PHASE DIAGRAMS

If the components are such that their mutual attraction, in terms of their electronic structure, is sufficiently large then their combination may result in the formation of a compound often intermetallic at a certain fixed composition. A typical diagram representing this is shown in Fig. 12.7. As per phase rule, on either side of the composition of these compound must lie a two phase region.

The ΔG_L^M at temperature T_L when every thing is liquid is shown in Fig. 12.7(b) where it is apparent that over the entire range of composition it is the liquid phase that is stabler than solid.

At the m.p. of compound C, *i.e.*, at T both the ΔG_L^M and ΔG_C^M have same value and hence with exist in equilibrium with each other. The composition range of stability of a compound will be between the two eutectic points E_1 and E_2 over which liquid phase shall extend. At a still lower temperature T_{S-L} which is much lower than m.p. of compound, it will separate out in the range of composition a-b beyond which on either side is liquid phase stable. In other words the free energy graphs should show miscibility gap between composition a-b. The tangent shall be common to ΔG_L^M and ΔG_C^M but there will be two solid-liquid equilibria on either side separated by a miscibility gap as discussed earlier.

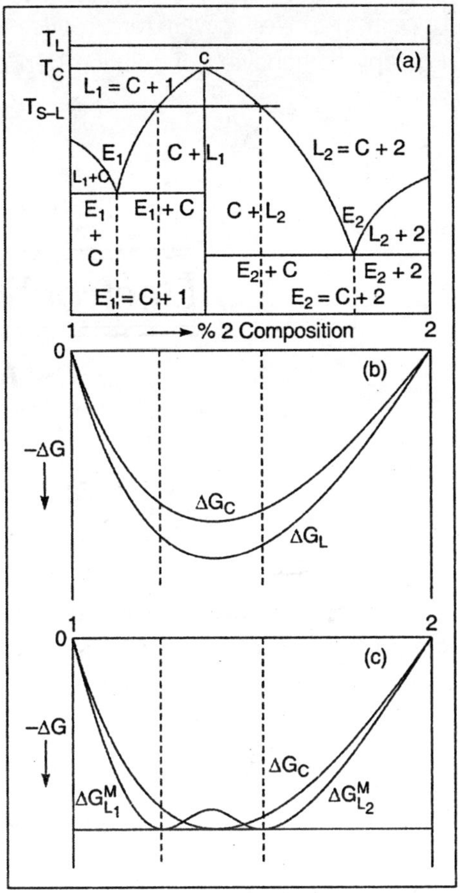

Fig. 12.7 Free energy-composition relationships showing the stabilities of intermetallic compounds.

12.7 CONCLUDING REMARKS

It is therefore clear that the phase diagrams are nothing but the depiction of equilibria between composition and temperature (pressure being kept constant) and the nature and stabilities of various phases can be guessed from their thermodynamic information like free energy of formation of the phases or the component activities and so on.

□□□

Chapter 13

Electrochemical Processes

Emf Measurements

13.1 INTRODUCTION

It is a known fact that compounds on dissolution in aqueous or fused media dissociate into respective lions. The conductivity of such a solution is due to the migration of these constituent ions. In such a solution if a metal (more precisely an element), whose ions exist in the solution, is dipped, the system now in the form of a metal in contact with its own ions, acquires a potential arising out of an equilibrium of a reaction like

$$M \rightarrow M^{n+} + ne \qquad ...(13.1)$$

between the metal and the solution. This system is known as galvanic element or an electrode. By virtue of the equilibrium of the above reaction, at the boundary of the metal and the solution, the metal part acquires a potential characteristic of its own and, dependent on concentration of ions (more precisely its chemical activity) and the temperature. This is known as the **'electrode potential'**. Two such elements or electrodes when joined as, electronic to electronic and ionic to ionic conductor, becomes a galvanic cell with potential as net algebraic sum of the two electrode potentials. Cell is formed because dissimilar metal electrodes have different potentials. Electrochemical cells are thus of two types, *viz.*, **concentration and galvanic cells**. Two dissimilar metal electrodes when joined make a galvanic cell and two similar metals but with different concentrations of the solutions around them, when connected, form a concentration cell. A solution in such cells is therefore referred to as electrolyte, A typical example of galvanic cell is Daniel Cell with $Cu/CuSO_4$ as one electrode and $Zn/ZnSO_4$ as another.

Because the electrode potential is due to the reaction as mentioned above, the overall cell reaction in Daniel Cell shall be

$$Zn + CuSO_4 = ZnSO_4 + Cu \qquad ...(13.2)$$

It means that Zn has a tendency to go in solution and eject Cu out of solution as stated above. Likewise in a concentration cell the natural process is that the meal ion from electrolyte of high concentration shall move to the lower concentration electrolyte.

13.2 CELL REACTION

In either case of these two cells as described above, as the ions move over a finite distance (greater than interatomic distance) electric current has to flow and thus the electrical circuit needs to be completed. This is possible in both these cases as shown in Fig. 13.1.

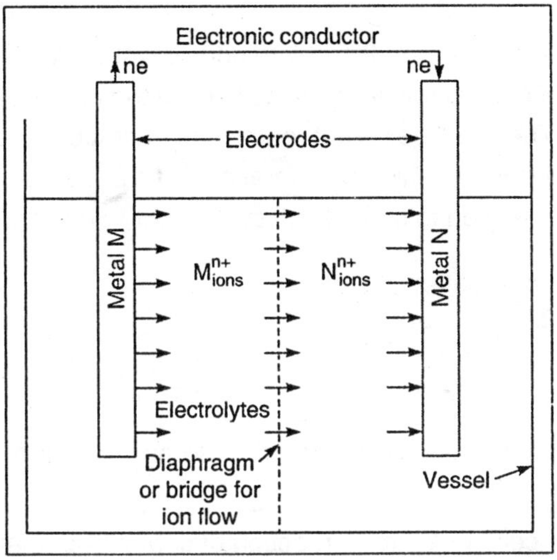

Fig. 13.1 Galvanic cell showing the ion and electron flows.

The electrode reaction as in Eq. (13.1) is an anodic dissolution reaction or is known as anode-half-cell reaction generating electrons which are transferred through the electronic conductor to the other electrode where the reaction

$$N^{n+} + ne \rightarrow N \qquad \text{...(13.3)}$$

takes place resulting in deposition of N on the cathode or it is the cathode-half-cell reaction which consumes electrons generated at he anode by the anode-half-cell reaction. The overall cell reaction therefore shall be

$$M + N^{n+} \rightarrow M^{n+} + N \qquad \text{...(13.4)}$$

There is a convention of describing a electrochemical cell. The Daniel Cell canbe represented as

$$\underset{(-ve)}{Zn} \mid ZnSO_{4\,(sol)} \parallel CuSO_{4\,(sol)} \mid \underset{(+ve)}{Cu}$$

in such a way that positive charge moves from left to right, *i.e.*, left hand electrode is negative (anode) and right hand electrode is positive (cathode). The vertical lines in the above represent the phase boundaries. The cell reaction is as shown in Eq. (13.2) and which is a natural process. It only means that Zn is at higher chemical potential than Cu or $CuSO_4$ is at higher chemical potential than $ZnSO_4$ and hence the above reaction is naturally possible. Consequently the cell shows *electro-motive force, emf* for short, across the two electrodes. The concentration cell can then be described as

$$Zn \mid \underset{(C_1)}{ZnSO_{4\,(sol)}} \parallel \underset{(C_2)}{ZnSO_{4\,(sol)}} \mid Zn$$

where concentration C_1 < concentration C_2. The natural process is that the higher concentration (C_2) solution will decrease and the lower concentration C_1 shall increase and accordingly the current shall flow. In this case by dissolving Zn on the left hand side the C_1 concentration of $ZnSO_4$ shall increase and by precipitating Zn out of $ZnSO_4$ from solution of concentration C_2 shall decrease.

If external emf is superimposed on the cell to counteract the cell emf then a stage can each when the superimposed emf exactly matches the cell emf in the opposite direction, when no current shall flow in the cell circuit. There will be no net reaction of migration of ions or electrons. This is the equilibrium state and all thermodynamic considerations of the cell, since at equilibrium or under reversible condition, can be applied to such a cell system with no net current flow.

■ **Example 13.1** The electrode potentials for

$$Co^{+2} \rightarrow Co^{+3} + e \qquad \text{is} - 1.82 \text{ volts}$$

and for

$$Sn^{+2} \rightarrow Sn^{+4} + 2e \qquad \text{is} - 0.15 \text{ volts.}$$

Which one of these two will reduce the other? Also establish the cell that will be useful as a source of emf.

Solution : Subtract the half cell or electrode reaction of cobalt from that of tin. It gives

$$Sn^{+2} + 2Co^{+3} = 2Co^{+2} + Sn^{+4}$$

the cell emf would be the algebraic sum of these two, *i.e.*,

$$\text{emf} = -(0.15) - (-1.82) = +1.67$$

as will be shown in next section the free energy change of such a electrochemical processes is

$$\Delta = -n\mathcal{EF} = 2 \times 123061 \times 11.67 = -77023 \text{ cal/g mole}$$

(wherein the value of $\mathcal{F}$ has been used as 23061 cal/v/g. equivalent)

This is thus a feasible process and hence Sn^{2+} shall reduce Co^{3+} or reaction of tin shall reduce reaction of cobalt. The cell therefore may be constructed as :

$$\text{inert metal} \mid Sn^{+2} \mid Sn^{+4} \text{ ions} \parallel Co^{+3} \text{ ions} \mid Co^{+2} \mid \text{inert metal}$$

wherein the positively charged ions shall move from left to right as is required and that this cell shall have net emf as 1.67 volts.

13.3 INTERCONVERSION OF FREE ENERGY AND ELECTRICAL WORK

Now,

$$G = H - TS$$
$$= U + PV - TS$$

On differentiation

$$dG = dU + P\,dV + V\,dP - T\,dS - S\,dT$$

where according to first law

$$dU = q - P\,dV - \partial w'$$

and substituting in this the second law statement gives

$$dU = T\,dS - P\,dV - \partial w'$$

therefore,

$$dG = V\,dP - S\,dT - \partial w'$$

When the cell works in open normal atmosphere at room temperature dp = 0 and dt = 0 and therefore,

$$dG = \partial w' \qquad ...(13.5)$$

The electrical work done by the passage of current, *i.e.*, electrical charge from one potential to another over a potential difference

$$w' = \text{charge} \times \text{potential difference (voltage)}$$

As per Farraday's Laws for cell reaction when one gram atom is deposited, one gram electron or 96500 coulombs of charge flows through. This is called **Farraday** and is denoted by '$\mathcal{F}$'. The number of electrons involved in a cell reaction, if n, then $n\mathcal{F}$ is the total electrical work done by the cell through the cell reaction and if the cell voltage or emf is $\mathcal{E}$ then the work done or free energy change for one mole is

$$\Delta G = -\, n\mathcal{E}\mathcal{F} \qquad ...(13.6)$$

If dn is the number moles of the ions passing through the cell for a spontaneous cell reaction then

$$\Delta G = -\, n\mathcal{E}\mathcal{F}\,dn \qquad ...(13.7)$$

Now, since $\Delta G°$ is the standard free energy change then corresponding $\mathcal{E}$ would be $\mathcal{E}°$, therefore,

$$\Delta G = -\, n\mathcal{E}°\,\mathcal{F} \qquad ...(13.8)$$

substituting in

$$\Delta G - \Delta G° = -\,RT \ln K \quad \text{as in Eq. (7.8)}$$

$$-\, n\mathcal{E}\mathcal{F} + n\mathcal{E}\mathcal{F}° = -\,RT \ln K$$

or

$$\mathcal{E} = \mathcal{E}° + \frac{RT}{n\mathcal{F}} \ln K \qquad ...(13.9)$$

The same relationship can be defined by considering a general reaction as in evolving the equilibrium constant wherein as in Eq. (7.6).

$$\Delta G - \Delta G° = RT \ln \frac{a_C^c\, a_D^d}{a_A^a\, a_B^b}$$

$$n\mathcal{E}\mathcal{F} - n\mathcal{E}°\,\mathcal{F}° = RT \ln \frac{a_C^c\, a_D^d}{a_A^a\, a_B^b} \qquad ...(13.10)$$

where $\mathcal{E}°$ is the cell emf when reactants and products are in there standard states and then

$$\Delta G° = n\mathcal{E}°\,\mathcal{F}$$

Thus,

$$\mathcal{E} = \mathcal{E}° + RT \ln \frac{a_C^c\, a_D^d}{a_A^a\, a_B^b} \qquad ...(13.11)$$

or

$$= \mathcal{E}° + RT \ln K \qquad \text{as in Eq. (13.9)}$$

13.4 ACTIVITIES AND EMF

Normally it is not possible to have a cell reaction in which several reactants and products will be constituting the cell reaction. These are very limited, often to one each and therefore,

$$\mathcal{E} = \mathcal{E}^\circ + RT \ln \frac{a_{product}}{a_{reactants}} \qquad ...(13.12)$$

This is also valid at only reasonable concentration and not at too low a concentration of either the product or reactant, otherwise $\mathcal{E}$ might tend to become infinite.

The concentration cell is therefore simple enough to analyse. It can be represented as

Pure metal A	electrolyte with A^{n+} ions	impure metal A *i.e.*, alloy

The cell reaction in this case is

$$A_{(pure)} \rightarrow A_{(impure\ or\ alloy)}$$

$$\Delta G = \Delta G^\circ + RT \ln \frac{a_{A\ (impure)}}{a_{A\ (pure)}} \qquad ...(13.13)$$

$a_{A\ (pure)}$ is one and when both are pure electrodes the cell emf is zero and hence $\Delta G^\circ = 0$ and it will reduce the above to

$$\Delta G = RT \ln a_A = -n_A \mathcal{E}\mathcal{F} \quad \text{or} \quad \log a_A = \frac{n_A \mathcal{E}\mathcal{F}}{4.575\ T}$$

which is extensively used for estimation of activities of various molten alloys using emf measurements.

Also activities of compounds in solution can also be found from galvanic cells. Let us take cell as

$$\text{W wire} \mid Ag \mid Ag_2O \,.\, B_2O_3 \,.\, SiO_2 \mid O_{2\,(1\ atmosphere)} \mid \text{Pt wire}$$

The net cell reaction is

$$4\,(Ag) + \{O_2\} = (2Ag_2O)_{1\,(electrolyte\ melt)}$$

$$\mathcal{E}_1 = \mathcal{E}^\circ + \frac{RT}{n\mathcal{F}} \ln a^2$$

similarly at some other concentration of (Ag_2O) the emf can be measured as

$$\mathcal{E}_1 - \mathcal{E}_2 = \frac{RT}{n\mathcal{F}} \ln \left(\frac{(Ag_2O)_1}{(Ag_2O)_2} \right)$$

Out of these $(Ag_2O)_2$ can be taken as pure or melt saturated with Ag_2O and therefore, the $(Ag_2O)_2$ shall be one and subsequently the activity of Ag_2O at other concentration can be found out. This technique in principle is used extensively for estimation of activities of oxide components in molten oxide systems of interest in extractive metallurgy and in refractory materials.

■ **Example 13.2** For the cell

$$Mo \mid Ag\text{-}Si \mid \underset{\text{(silica sat. melt)}}{B_2O_3 - SiO_2} \mid O_{2\,(1\ atmos)} \mid Pt$$

the cell emf was 1.70 volts with Ag-Si as saturated with silicon. The cell emf was 1.55 volts when Ni_{Si} = 0.1 in the Ag-Si alloy at 11150°C. Also calculate the standard free energy of formation of silica at 1100°C and activity coefficient of silicon at N_{Si} = 0.1.

Solution : The cell reaction is $Si + O_2 = SiO_2$ where all are in their standard states. Therefore,

$$\Delta G° = -n\mathcal{E}\mathcal{F} = -4 \times 1.7 \times 123061 = -836814 \text{ cal/g mole}$$

(wherein the value of Farraday constant is 23061 cal/v/equiv)

When silicon is not at unit activity then the cell reaction can be written as

$$[Si] + \{O_2\} = \langle SiO_2 \rangle$$

and

$$K = a \langle SiO_2 \rangle / a_{[Si]} \times p_{\{O_2\}}$$

and since the activity of silica is unity because the electrolyte is saturated with silica and since oxygen is at one atmosphere therefore,

$$K = 1/a_{[Si]}$$

Also the non-standard free energy of formation of silica is

$$\Delta G = -n\mathcal{E}\mathcal{F} = -4 \times 1.55 \times 123061 = -762978$$

Hence,

$$\Delta G = \Delta G° + RT \ln K$$

or

$$-142978 = -156814 + RT \ln \frac{1}{a_{[Si]}}$$

or

$$RT \ln \frac{1}{a_{[Si]}} = -762978 - 836814 = -73836$$

or

$$\log a_{[Si]} = 2.11, \textit{ i.e., } a_{[Si]} = 0.007$$

or

$$\gamma_{Si} = a_{[Si]}/N_{[Si]} = \frac{0.007}{0.1} = 0.07$$

■ **Example 13.3** Calculate the free energy change at 800°C for the reaction

$$FeO + Ni = NiO + Fe$$

if the cell emf for the cell

$Mo \mid MoO_2 \mid electrolyte \mid FeO \mid Fe$ is 17.8 mV

and for cell $Mo \mid MoO_2 \mid electrolyte \mid NiO \mid Ni$ is 28.4 mV, both at 800°C.

Solution : Cell reaction of $Mo \mid MoO_2 \mid electrolyte \mid FeO \mid Fe$

is

$$Mo + 2FeO = MoO_2 + 2Fe$$

and for which the free energy change is, assuming FeO in pure form

$$\Delta G^{\circ}_{Fe} = -n\mathcal{E}\mathcal{F} = -RT \ln \frac{a_{MoO_2} \times a^{a}_{Fe}}{a_{Mo} \times a^{2}_{FeO}}$$

and for the cell $Mo \mid MoO \mid electrolyte \mid NiO \mid Ni$

cell reaction is $$Mo + 2NiO = MoO_2 + 2Ni$$

and $$\Delta G^\circ_{Ni} = -n\mathcal{EF} = -RT \ln \frac{a_{MoO_2} \times a^2_{Fe}}{a_{Mo} \times a^2_{NiO}}$$

For the reaction $$FeO + Ni = NiO + Fe$$

the free energy change $$\Delta G^\circ_{Fe\text{-}Ni} = -RT \ln \frac{a_{NiO} \times a_{Fe}}{a_{Ni} \times a_{FeO}}$$

Dividing $$\Delta G_{Fe\text{-}Ni} = \Delta G^\circ_{Fe} / \Delta G^\circ_{Ni} = 2 \ln \frac{a_{NiO} \times a_{Fe}}{a_{Ni} \times a_{FeO}} = \frac{4 \times 17.8 \times \mathcal{F}}{4 \times 28.4 \times \mathcal{F}}$$

Therefore, $$\Delta G^\circ = -RT \ln = \frac{17.8}{28.4}$$

$$= -650 \text{ cal/g mole}$$

13.5 TEMPERATURE COEFFICIENT OF CELL EMF

Now, since $$\Delta G = -n\mathcal{EF}$$

$$d\,\Delta G = -n\mathcal{F}\,d\mathcal{E}$$

$$\left(\frac{d\,\Delta G}{dT_P}\right) = -n\mathcal{F}\frac{d\mathcal{E}}{dT}$$

and at constant pressure

$$\left(\frac{d\,\Delta G}{dT}\right)_P = -\Delta S$$

and since cells are generally operated at atmospheric pressure this condition holds true.

Therefore, $$\Delta S = n\mathcal{F}\frac{d\mathcal{E}}{dT}$$

$$= -n\mathcal{EF} + n\mathcal{F}T\frac{d\mathcal{E}}{dT}$$

Once ΔG, ΔH and ΔS are known along with the activities of the constituent of the electrode then by using the following

$$\overline{G}^M_A = RT \ln a_A = -n_A\,\mathcal{EF}$$

$$\overline{S}^M_A = n_A\,\mathcal{F}\left(\frac{d\mathcal{E}}{dT}\right)_P$$

$$\overline{H}^M_A = n_A\,\mathcal{FE}\left[\mathcal{E} - T\left(\frac{d\mathcal{E}}{dT}\right)_P\right]$$

and excess properties can also be calculated.

■ **Example 13.4** The emf of the cell Ce | $CdCl_2$-$CuCl_2$ | Cu is 1.2 volts at 0°C and 1.15 volts at 25°C. Calculate the ΔH of the cell reaction at 25°C.

Solution : The anode reaction is $$Cd \rightarrow Cd^{+2} + 2e$$

and the cathode reaction is $$Cu^{+2} + 2e \rightarrow Cu$$

The free energy change of the cell reaction is

$$\Delta G_{273°K} = -2 \times 1.2 \times 23061 = -55346$$

and

$$\Delta G_{298°K} = -2 \times 1.15 \times 23061 = -53040$$

$$-\Delta S = \frac{\partial \Delta G}{\partial T} = \frac{\partial \Delta(298) - \Delta G(273)}{298 - 273}$$

and after putting the above values $= 92.2$ cal/g mole/0°

Now, since

$$\Delta H = \Delta G - T\,\Delta S$$
$$= -53040 - 298 \times (-92.2)$$
$$= -25624 \text{ cal/g. mole}$$

13.6 SOLID ELECTROLYTE CELLS

Certain solid solutions behave as ionic solutions since the current is carried through them by conduction of ions. Some typical examples are ZrO-CaO(4%), ThO_2-CaO, ThO_2-Y_2O_3(7%) or ThO_2–Lanthana. These behave as ionic conductors above a certain oxygen partial pressure which is fairly low like 10^{-10} atms at 1000°C. Since these are oxide cells made with special oxides as electrolytes at high temperatures, these can be and have been used for thermodynamic measurements of oxidation reduction reactions at various pressures of oxygen including vacuum.

The cell used are of the types :

$$<Cr> - <Cr_2O_3> \mid ZrO_2 - (4\%\ CaO) \mid [O] \text{ in Fe alloy}$$

$$<Cr> - <Cr_3O_3> \mid ZrO_2 - (7\%\ Y_2O_3) \mid [O] \text{ in Fe alloy}$$

for which the cell reaction is

$$\frac{2}{3} <Cr> + [O] = -\frac{1}{3} <Cr_2O_3>$$

or

$$\mathcal{E} - \mathcal{E}° = \frac{RT}{n\mathcal{F}} \ln a_O$$

since both Cr and Cr_2O_3 are in solid state.

Where $\mathcal{E}°$ is calculated from the measurements of cell when [O] in Fe is replaced by oxygen at one atmospheric pressure or can be calculated knowing the free energy of formation of $<Cr_2O_3>$ taking $\mathcal{E}$ as 23061 cal/v. This has been reduced to

$$\log a_O = 4.62 - \frac{13580 - 10.08\,\mathcal{E}}{T}$$

where $\mathcal{E}$ is expressed in millivolts and T in K.

By employing the right hand electrode as [O] in Fe with any other element like C, P, Mn, S etc. the

$$\log f_O^i = e_O^i\ (\%\ i)$$

and thus even the interaction parameters can also be found out. The results have been extremely good in that the calculated ones tally with actual measured ones in most cases.

A cell of the type

<Mo> | <MoO_2> | ZrO_2–CaO | <Ni>, <NiO>

if set up, means

$$\mathcal{E} = \mathcal{E}^\circ + RT \ln \frac{a_{MoO} \cdot a_{Ni}^2}{a_{NiO}^2 \cdot a_{Mo}}$$

$$\mathcal{E} = \frac{\text{Chemical potential of Ni-NiO} - \text{Chemical potential of Mo-MoO}_2}{n\mathcal{E}}$$

Knowing chemical potential, *i.e.*, partial molar free energy of one oxide the other can be found out by measuring the cell emf $\mathcal{E}$. Of alternatively cell emf $\mathcal{E}$ is related to the free energy of cell reaction and which is the algebraic sum of the free energies of products minus reactants and from these the ΔG of one oxide can be found out knowing ΔG of the other oxide.

13.7 CONCLUDING REMARKS

Electrochemical measurements are many. For example the conductance measurements, ion transport estimation and so on. But the electrochemical process or the change, is only the change in chemistry accompanied by the passage of current through the system. The energy exchange is inalienably associated with this process and hence is of direct interest in thermodynamic measurements. The technique of emf measurement has been extensively exploited for the estimation of thermodynamic parameters, in particular the activities and therefrom the other energy properties of the system in which the chemical reactions occur in the form of electrochemical processes. The use of solid electrolytes has opened up a large vista for measurements at fairly high temperatures.

□□□

Chapter 14

Free Energy-Temperature Relationships

Ellingham Diagrams

14.1 INTRODUCTION

The free energy composition relationships were discussed in Chapter 12 to conform the nature of phase equilibria involved in phase diagrams. Often it is necessary to know the free energy-temperature relationships, which are nothing but equilibria as affected by temperature, to establish the temperature at which it is advisable to carry out a certain process, particularly the chemical processes.

Free energy change in a process is an indication of whether the process is at equilibrium, or it is natural or spontaneously occurring or, it is an unnatural process. Since most processes of interest are as such unnatural processes (see Section 1.1), the conditions, particularly the temperature is of direct interest for its commercial exploitation. **Free energy-temperature-pressure relationships** are of direct use in this regard.

Any material form, say for example iron oxide, as some oxide mineral of iron, needs to be reduced to produce iron. Iron oxide as such has a fairly low value of free energy and hence stable. Had it not been so iron would have been produced as a natural process, which is not the case. Conditions can be altered suitably such that iron oxide becomes unstable as compared to something else. For doing this pure iron oxide system is transformed into an iron oxide plus carbon system and heated to a suitable temperature (above 900°C). The free energy of this mixture is higher than that of iron and whatever other products, CO or CO_2 produced at the high temperature. The system therefore undergoes a transformation and produces iron as desired. It means that reaction

$$Fe_2O_3 + 3C = 2Fe + 3CO \qquad ...(14.1)$$

takes place and the reduction is possible to produce iron. Knowledge of free energy-temperature relationship is guidance to convert otherwise stable system into a unstable one and get the desired results.

In the reaction above, original iron oxide is destroyed, and a different one, oxide of carbon is produced. The condition of temperature, is particular, is fixed in such a way that the above reaction is possible or net free energy change is negative for the overall reaction. The above reaction consists of two reaction equilibria as

$$2Fe + 3/2\ O_2 = Fe_2O_3 \quad ...(14.2)$$

$$2C + O_2 = 2CO \quad ...(14.3)$$

which are, manipulated suitably for reaction in Eq. (14.1) to occur.

With reference to above it can be said that if the relative stabilities of oxides, as they vary with temperature, are known a suitable pair can be worked out as above to carry out the desired process. The qualitative term stability is quantified as free energy associated with these various oxides. More negative is the free energy of formation of the oxide the more stable is the oxide.

The same arguments can be made for any group of substance like sulphides, sulphates, chlorides, silicates and so on, in which case one sulphide will be destroyed to produce another sulphide and the like. It means for example, that

$$Sb_2S_3 + 3Fe = 2Sb + 3FeS \quad ...(14.4)$$

would be possible to produce antimony provided the free energy of formation of iron sulphide is more negative than that for antimony sulphide at some temperature at which the reaction will have to carried out.

All chemical processes are essentially making, breaking or both of chemical bonds. In the manufacturing processes chemical compounds are broken and new ones are formed. Being non-nuclear, the element remains the same before and after reaction but the compounds change from reactants to products.

The free energy of compounds, *i.e.,* free energy of formation of compounds can act as a stability scale for compounds and since free energy is a function of temperature the stability shall also be subject to temperature as a variable. The free energy of formation of CO is more negative than that for iron oxide and hence reaction as in Eq. (14.1) is readily possible. Similar is the story of reaction in Eq. (14.4).

The free energy value changes with associated interacting substance, *i.e.*, with the activity of the constituents in reactants and products as

$$G = G° + RT \ln K$$

where,

$$K = \frac{\text{product of activity of products}}{\text{product of activity of reactants}}$$

Hence, the changeover from pure to impure form of reactants of products are going to affected the free energy and hereby the relative stabilities.

All these factors are quantitatively examined in the following in terms of free energy-temperature relationship.

14.2 FREE ENERGY OF REACTIONS INVOLVING ONLY GASES

This group is separately classified because of its characteristic difference from those reactions involving condensed phases. These reactions can be typically represented by

$$2H_2 + O_2 = 2H_2O \quad ...(14.5)$$

$$2CO + O_2 = 2CO_2 \quad ...(14.6)$$

$$2SO_2 + O_2 = 2SO_3 \quad ...(14.7)$$

Let us consider the free energy of reaction (14.5)

$$\Delta G = 2G_{H_2O} - 2G_{H_2} - G_{O_2}$$

This will be in equilibrium if $\Delta G = 0$

Therefore, $2G_{H_2O} = 2G_{H_2} - G_{O_2}$

Putting, $G_i = G_i^\circ + RT \ln p_i$ as in Eq. (4.56)

$$2G^\circ_{H_2O} + 2RT \ln p_{H_2O} = 2G^\circ_{H_2} + 2RT \ln p_{H_2} + G^\circ_{O_2} + RT \ln p_{O_2} \quad ...(14.8)$$

where p_{H_2O}, p_{H} and p_{O_2} are corresponding partial pressures of the components involved in the reaction.

Since, $\Delta G^\circ_{(14.5)} = 2G^\circ_{H_2O} - 2G^\circ_{H_2} - G^\circ_{O_2}$

therefore, $\Delta G_{(14.5)} = 2RT \ln p_{H_2} + RT \ln p_{O_2} + 2RT \ln p_{H_2O} \quad ...(14.9)$

$$= 2RT \ln \frac{p_{H_2} \cdot p_{O_2}^{1/2}}{p_{H_2O}}$$

$$= - RT \ln K_P$$

or $$2 \log K_P = \frac{-\Delta G^\circ}{4.575\ T} \quad ...(14.10)$$

where K_P is the equilibrium constant of the reaction when gases alone are involved. It is different from the equilibrium constant K_X which takes into account condensed phases as discussed in Sec. 7.2.

For reaction (14.5) it has been estimated

$$\Delta G^\circ_{(14.5)} = -118000 - 26.75\ T$$

therefore at 1600°C or 1873°K

$$2 \log K_{P(14.5)} = + \frac{118000}{4.575\ T} + \frac{26.75\ T}{4.575\ T} = + 19.616$$

or $\log K_P = 9.808$

If it is assumed that the atmosphere required should have

$$p_{O_2} = 10^{-4} \text{ atmos}$$

then $$\log K_P = 9.808 = \frac{p_{H_2} \times 10^{-2}}{p_{H_2O}}$$

or $$\frac{p_{H_2}}{p_{H_2O}} 9.808 \times 10^2 = 980$$

If p_{H_2} is one atmosphere then p_{H_2O} = 1/980 = say approximately 1 mm Hg.

The temperature of water when its water vapour has a pressure of 1 mm Hg and if hydrogen gas is bubbled through the water at this temperature then the resulting gas mixture shall have a ratio of p_{H_2}/p_{H_2O} as desired above to be equivalent to partial pressure of oxygen equal to 10^{-4} atmosphere. Similarly, for reaction (14.6)

$$\Delta G^\circ = -115200 + 41.50\,T$$

$$2 \log K_P = +\frac{115200}{4.575\,T} + \frac{41.50}{4.575}$$

at 1273°
$$2 \log K_P = 19.78 - 9.071 = 28.85$$

or
$$\log K_P = 14.42$$

or
$$\log K_P = \frac{p_{CO}}{p_{CO_2}} \times p_{O_2}^{1/2} = 14.42$$

If partial pressure of oxygen is to be kept at $p_{O_2} = 10^{-2}$ atmos.

$$\frac{p_{CO}}{p_{CO_2}} = 14.42 \times 110 \text{ atmos.} \quad \text{or} \quad \frac{p_{CO}}{p_{CO_2}} = 142$$

The CO and CO_2 gases will have to be mixed in such a way as to obtain the above ratio to obtain $p_{O_2} = 10^{-2}$ atmos. It means 142 ml of CO will have to mixed with 1 ml of CO_2 to get the required mixture. Often CO_2 is generated and passed over carbon at appropriate temperature when $CO_2 + CO = 2CO$ reaction shall occur and the appropriate ratio of CO/CO_2 will be obtained. The details shall be discussed later.

It is thus clear that if the sum total of powers to which each of the reactants and products are to be raised are equal, the equilibrium shall be independent of pressure but if these two are not equal, which shall often be the case, these equilibria involving only gaseous components shall be dependent on pressure. The system shall then be defined at a given temperature and pressure and in that case at equilibrium the free energy change will be zero.

These reactions are often between two elemental component gases giving rise to a gaseous compound and often shall be pressure dependent. A reaction of the type

$$CO + H_2O = CO_2 + H_2 \qquad \text{...(14.11)}$$

can be imagined and since

$$K_P = \frac{p_{CO_2} \times p_{H_2}}{p_{CO} \times p_{H_2O}}$$

appears as pressure independent. A close scrutiny of such a reaction shall reveal that this reaction is a combination of reactions in Eqs. (14.5) and (14.6). It only means that the two shall be independently in equilibrium at a fixed p_{O_2} value. For the Eq. (14.1) to be in equilibrium, p_{O_2} for both of the reactions in Eqs. (14.5) and (14.6) shall have to be equal. In other words the pressure is fixed for the equilibrium and that is what the K for (14.11) reveals.

14.2.1 Oxygen Potential-Temperature Relationships

Eq. (14.5) gives

$$\Delta G^\circ = -RT \ln K_P = RT \ln\left(\frac{p_{H_2}^2}{p_{H_2O}^2} \times p_{O_2}\right) \qquad \text{...(14.12)}$$

or $$RT \ln p_{O_2} = \Delta G° - 2RT \ln\left(\frac{p_{H_2}}{p_{H_2O}}\right) \quad ...(14.13)$$

For the reaction (14.5)

$$\Delta G° = -\ 118000 + 26.75\ T \text{ cal/mole}$$

Therefore, $$RT \ln p_{O_2} = -\ 118000 + 26.75\ T - 2RT \ln\left(\frac{p_{H_2}}{p_{H_2O}}\right) \quad ...(14.13a)$$

For any fixed ratio of H_2/H_2O the RT ln p_{O_2} can be correlated with temperature and it shall be a straight line. The intercept of the line on y-axis, *i.e.*, RT ln p_{O_2} axis shall be – 118000 cal. This is shown in Fig. 14.1 in the form of point 'H'. When H_2/H_2O ratio is one the RT ln p_{O_2} shall increase (move in less negative direction because of +26.75 T factor) with temperature and hence the line is sloping upwards from the anchor point 'H' as temperature rises. If H_2/H_2O ratio increase beyond one, for every ratio the line will slope further up but originating from 'H'. If H_2/H_2O is less than one \the term 2RT ln (p_{H_2}/p_{H_2O}) shall be +ve and hence the RT ln p_{O_2} will move to more –ve side but every time the point H shall remain the same for all such relationships and hence all lines for various H_2/H_2O ratios shall originate from point 'H'. For all RT ln p_{O_2} has the same value as shown by point H. As H_2/H_2O ratio decreases below one the lines shall rotate in clockwise and when it increases beyond one in anti-clockwise direction as shown in Fig. 14.1.

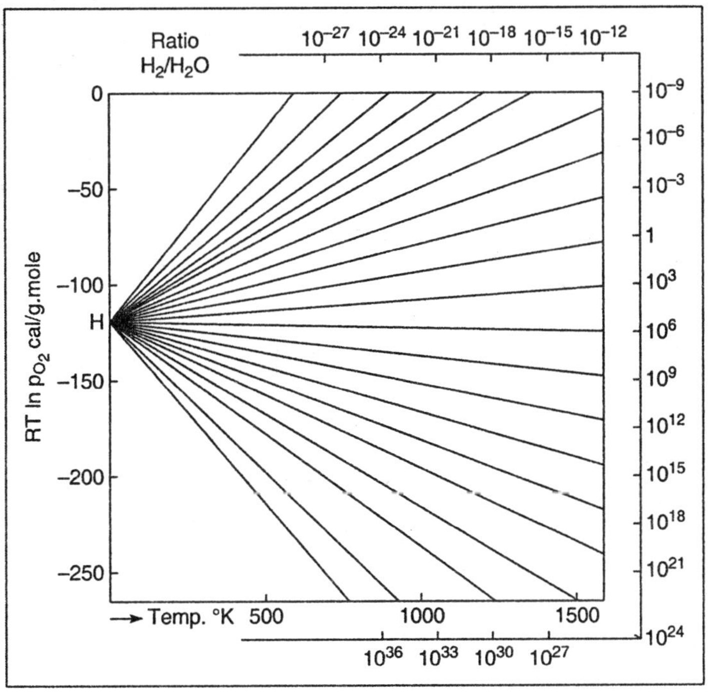

Fig. 14.1 Variation of RT ln p_{O_2} of the reaction $2H + O_2 = 2H_2O$ with temperature.

Additional scale of H_2/H_2O ratio is superimposed on this diagram in the form of additional y-axis. In order to make the diagram a compact one since the y-axis extends on either side of x-axis over a long range, the y-axis of H_2/H_2O ratio is bent at right angles and made parallel with x-axis.

The lines emanating from the point H upto the vertical y-axis shall cut across the bent y-axis and which is now marked by corresponding values of the gas ratio of H_2/H_2O. The obvious the need to have long y-axis and thereby the diagram becomes more compact.

For the reaction in Eq. (14.6) the free energy change

$$\Delta G^\circ = -135000 + 41.57\ T\ \text{cal/mole.}$$

This can be processed as was done in the case of reaction (14.5) to get

$$RT \ln p_{O_2} = -135000 + 41.57\ T - RT \ln\left(\frac{p_{CO}}{p_{CO_2}}\right) \qquad ...(14.15)$$

and by the same logic RT ln p_{O_2} can be plotted against temperature for fixed values of CO/CO_2 ratios as shown in Fig. 14.2.

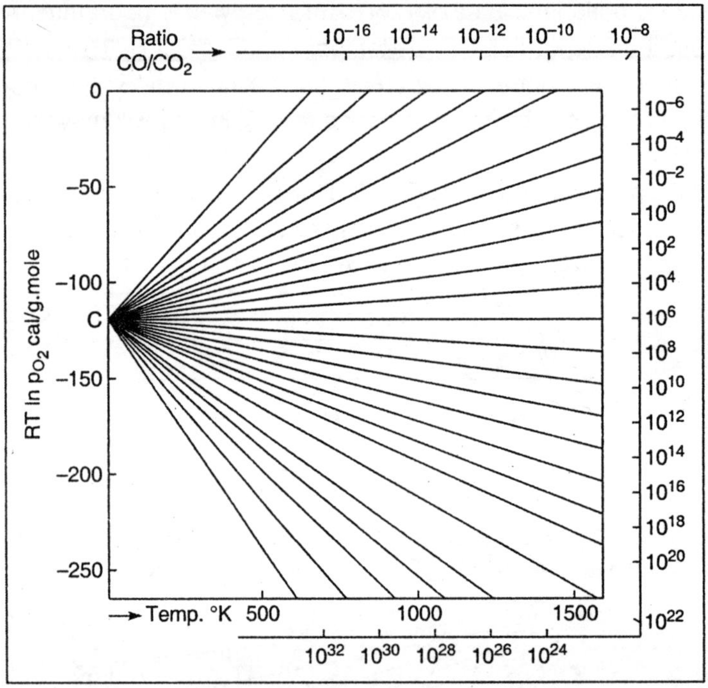

Fig. 14.2 Variation of RT ln p_{O_2} of the reaction $2CO + O_2 = 2CO_2$ with temperature.

The value of RT ln p_{O_2} at zero degree absolute is −135000 cal and is shown as part 'C'. All the lines for various CO/CO_2 ratios shall originate from point 'C' and shall move clockwise on decreasing the ratio below one and anti-clockwise beyond ratio one on increasing the CO/CO_2 ratio. In this diagram the additional y-axis of CO/CO_2 ratio and is shown in the same way as in Fig. 14.1.

This is the standardised treatment and may apply to even reactions in Eq. (14.7) likewise. Similarly, for a reaction

$$2H_2 + S_2 = 2H_2S$$

the RT ln p_{O_2} shall now be replaced by RT ln p_{S_2} and related to temperature at fixed ratios of H_2/H_2S. This is useful for sulphidation reactions. Even a reaction of the type

$$2H_2 + Cl_2 = 2HCl$$

could be amenable for chlorination processes as a function of temperature and H_2/HCl ratios.

So far the ratios H_2/H_2O, CO/CO_2 and the like are used to work out their variation as a function of temperature and a function RT ln p_{O_2}. The requirement is however to know the variation of Rt ln p_{O_2} as such as a function of temperature and partial pressure of oxygen itself.

This could be established using the relation given in Eq. (4.56) as

$$G = G^\circ + RT \ln p$$

If p_i is one atmosphere of pure oxygen and assuming it to be the standard state for oxygen then

$$G = G^\circ$$

confirming that It is the standard state and in equilibrium with one atmosphere of oxygen, which is obvious enough. The RT ln p_{O_2} = 0 shall be the correlation of RT ln p_{O_2} with temperature at p_{O_2} = 1 as shown in Fig. 14.3 in the form of the horizontal line passing through the point RT ln p_{O_2} = 0 at any temperature.

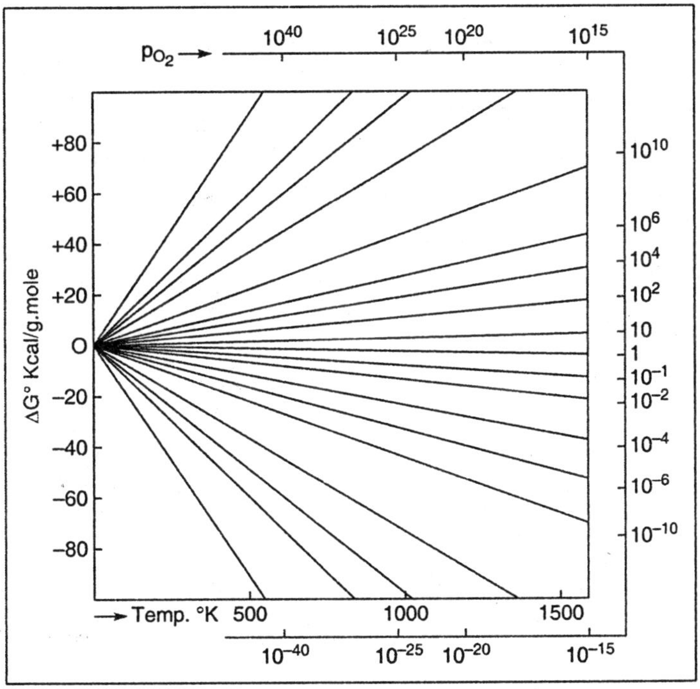

Fig. 14.3 Variation of RT ln p_{O_2} with temperature for various pressures of oxygen.

If the gas is now at a pressure less or more than one atmosphere then,

$$Rt \ln p_{O_2} = G - G^\circ$$

or in another words it will be the free energy change from standard state when p_{O_2} = 1 of a new state when $p_{O_2} \neq 1$, *i.e.*, for the reaction

Oxygen (one atoms) = oxygen (other the one atmos)

This free energy change varies with temperature for a fixed partial pressure of oxygen as can be seen in Fig. 14.3. All these lines shall originate from RT ln p_{O_2} = 0 at T = 0°K shown as point '0' in Fig. 14.3. If oxygen pressure is less than one atmos ($p_{O_2} < 1$) the lines shall move in towards more −ve free energy and if oxygen pressure is more than one atmos ($p_{O_2} > 1$) towards +ve free energy values. This is obvious because pure oxygen, being at higher chemical potential shall change to oxygen at less than one atmosphere as a natural process or spontaneous change with –ve free energy change and just the opposite when $p_{O_2} > 1$ atmos.

As before an additional y-axis as p_{O_2} is provided on the diagram as shown in Fig. 14.3.

The equilibrium of reactions as in Eqs. (14.5) and (14.6) indicate that at a fixed temperature if the CO/CO_2 or H_2/H_2O ratio is fixed then RT ln p_{O_2}, *i.e.*, p_{O_2} as such is fixed and the system is invariant.

In fact therefore, CO/CO_2, **H_2/H_2O and p_{O_2} are one and the same thing expressed in different forms for a certain fixed temperature.** This will be clear from the following.

Let us draw a few lines of different CO/CO_2 ratios originating from point C and different lines originating from point O such that they intersect each other as shown in Fig. 15.4. Let us draw a line originating from point H and passing through the above mentioned intersection.

Let us say that p_{O_2} and CO/CO_2 lines intersect at a point shown as A. Let us also work out the temperature corresponding to this point A as T as shown. Therefore, at temperature T the point A indicates certain fixed CO/CO_2 ratio to be read from relevant y-axis.

The same point A also indicate a certain fixed value of p in the direction O-A to be read from appropriate y-axis.

If a line is drawn from H through point A it will read the corresponding equilibrium H_2/H_2O ratio on the corresponding y-axis.

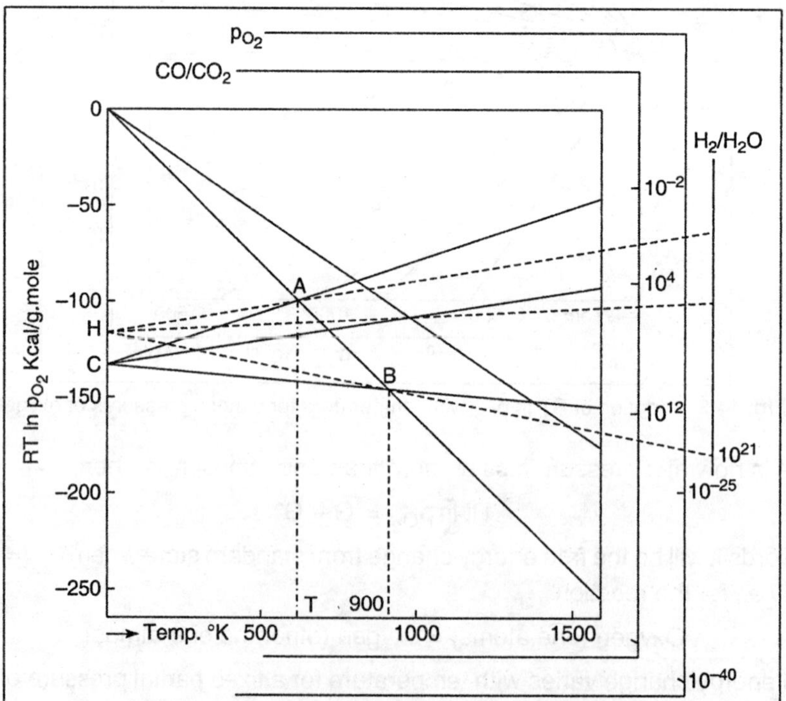

Fig. 14.4 The interrelationships of p_{O_2}, CO/CO_2 and H_2/H_2O ratios and the method of determining these graphically.

Therefore, at the temperature T in Fig. 14.4,

$$CO/CO_2 \text{ ratio of } 10^2 = p_{O_2} \text{ of } 10^{-40} = H_2/H_2O \text{ ratio of } 10^{-12}$$

In other words the oxygen potential can also be depicted at any temperature by corresponding ratio of CO/CO_2 or H_2/H_2O.

So for a certain p_{O_2} value at a certain temperature the equilibrium CO/CO_2 ratio can be found out like this. Draw the relevant p_{O_2} line originating from the point O. Draw a vertical through the required temperature to intersect this line. Draw a line from point 'C' through this intersection to read on CO/CO_2 ratio axis. The value obtained is required CO/CO_2 ratio. If instead of through C, line is drawn through point H it will read the H_2/H_2O ratio on the H_2/H_2O axis in equilibrium with p_{O_2}. This is shown in Fig. 14.4 by the point B at temperature of 900°K when p_{O_2} of 10^{-40} = CO/CO_2 ratio of 10^{-12} = H_2/H_2O ratio of 10^{11}. It is rather easier to obtain the equivalent CO/CO_2 or H_2/H_2O ratios for a given partial pressure of oxygen. All these indicate the oxygen potential of the system in various forms.

14.3 THREE ENERGY OF REACTIONS INVOLVING ONLY CONDENSED PHASES

This is other extreme when no gaseous component is involved in the process. Such reactions can be exemplified as

$$2CaO + SiO_2 = 2CaO \,.\, SiO_2 \qquad ...(14.12)$$

$$Al_2O_3 + 2CaO = 2CaO \,.\, Al_2O_3 \qquad ...(14.13)$$

or even carbide forming reactions from carbon and meal also falls in this category; and such others which are more common in ceramic systems. The reactions of this type are possible only because the two condensed phases have opposite chemical characters. Reactions of the type

$$CaO + MgO = \text{solid or liquid solution}$$

leads to solid or liquid solution and not any chemical compound. Out of the simple compound types of therefore, acidic oxides like SiO_2, P_2O_5 and sometimes Al_2O_3 form chemical compounds with alkali or alkaline earth metal oxides as silicates, phosphate and aluminates.

The free energy change of these reactions are independent of pressures. These however do change with temperature, as it should be. The free energy is given by

$$\Delta G° = \Delta H° - T\,\Delta S°$$

In which the enthalpy can be treated as substantially constant and is equal to the intercept of the line G vs. T on the G axis at zero degree absolute. The entropy change in a reaction of the type (14.12) shall be

$$\Delta S^{\circ}_{(14.12)} = S^{\circ}_{2CaO.SiO_2} - S^{\circ}_{2CaO} - S^{\circ}_{SiO_2}$$

Since all are condensed phases the individual entropy values are small and are often of the same magnitude for many substances in the same physical state. Therefore, ΔS° is very small for such reactions. The net result is that the ΔG° vs T plot for such reactions do not vary much with temperature and hence are parallel to the temperature axis as shown in Fig. 14.5. These lines do

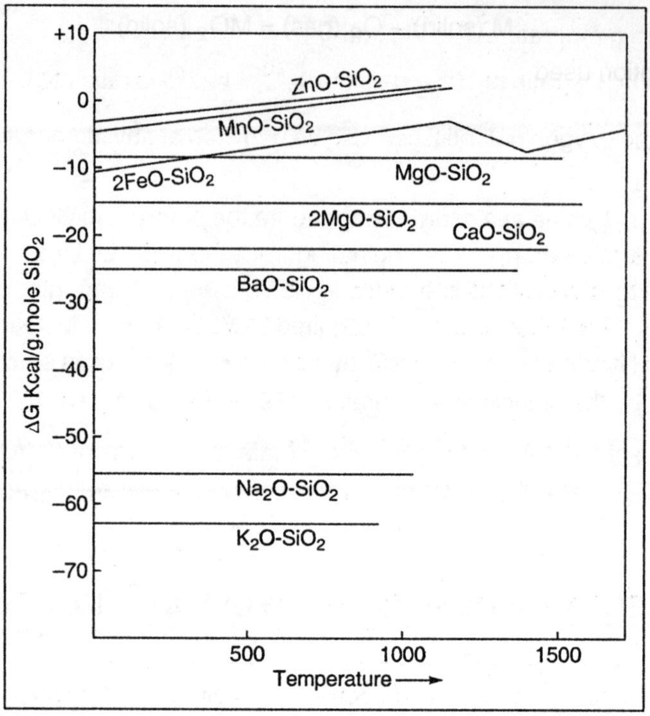

Fig. 14.5 Free energy-temperature relationships of reactions involving only condensed phases.

change their slopes if the reactants or products are not in their standard states. This non-standard behaviour shall be discussed in a generalised way later. These plots also change their slope when phase change occurs in either the reactants or the products or both. This also will be discussed in a generalised way for all reactions later.

14.4 FREE ENERGY OF REACTIONS INVOLVING CONDENSED AND GASEOUS PHASES

What will be the difference if condensed and gaseous components are involved in a reaction? There is considerable difference between the earlier two categories and this category as far as the free energy of reactions are concerned. By far this is the most widely encountered practical system, because chemical changes often involves a gaseous component like oxygen, sulphur, halide, carbon dioxide, carbon monoxide, and so on. In fact in the inorganic world these reagents dominate and hence become a part of processes for consideration. The free energies of reactions involving any one of these shall be a function of pressure of such gaseous components. The free energy does vary with the purity of the condensed phase(s) in the reaction but this deviation shall be discussed later. Initially only pure condensed phases shall be dealt with. The equilibria is typically described by reactions of the type :

Metal M + O_2 gas = MO_2 oxide

or Metal M + S_2 Gas = MS_2 sulphide and son on

Let us assume that the temperature for consideration to be below the melting point of both metal as well as the oxide and hence let us study the heterogenous reaction of interest exemplified as

$$M \text{ (solid)} + O_2 \text{ (gas)} = MO_2 \text{ (solid)} \qquad \ldots(14.14)$$

or as per the convention used

$$<M> + \{O_2\} = <MO_2> \qquad \ldots(14.14a)$$

The temperatures are such that the partial pressures of M and MO_2 in gas phase shall be extremely small and each of this solid phase shall be in equilibrium with its own vapour pressure. The free energy of solid phase is thus equal to the corresponding partial part in the gases phase. Under the present conditions the partial pressure of solids in a gas phase vary negligibly even when pressure changes considerably. Hence the free energy of solid phase can be conveniently assumed to be independent of pressure and thereby assumed to be constant. The gas phase equilibrium can be written in the form of reaction as in Eq. (14.14) and the solid states themselves to be taken as standard states.

The free energy change of reaction (14.14) is thus

$$\Delta G^\circ_{(14.14)} = \Delta G^\circ_{MO_2} - \Delta G^\circ_{M_2} - \Delta G^\circ_{O_2}$$

$$= -RT \ln \left[\frac{a_{MO_2}}{a_M \times p_{O_2}} \right] \qquad \ldots(14.15)$$

and $$\Delta G^\circ = -RT \ln K = +RT \ln p_{O_2} \qquad \ldots(14.16)$$

where K is the equilibrium constant of the reaction.

It also means that free energy can be expressed in terms of partial pressure of oxygen as

$$\Delta G^\circ = RT \ln p_{O_2} \qquad \ldots(14.17)$$

and therefore, they-axis plotted in Fig. (14.1), (14.2) and (14.3) as RT ln p_{O_2} is nothing but the standard free energy change ΔG° of the reaction. Thus the RT ln p_{O_2} vs. temperature plots for metal/metal oxide systems are nothing but the free energy/temperature relationships. Likewise the RT ln p_{O_2} shall be the free energy term for sulphides in such diagrams.

The $\Delta G^\circ_{(14.14)}$ is thus the standard free energy change of the reaction as in Eq. (14.14). Since K is a function of temperature, at any fixed temperature the equilibrium shall be decided by the unique value of the partial pressure of oxygen in contact with solid metal and its oxide. From phase rule it can also be deduced that the equilibrium in reaction (14.14) has only one degree of freedom.

Therefore, in a metal-metal oxide-oxygen system if partial pressure of oxygen is less than the equilibrium value, meal oxide will spontaneously reduce to make up the partial pressure of oxygen to the equilibrium value and if the oxygen partial pressure is more than the equilibrium, the excess oxygen will spontaneously read with metal to reduce itself to the equilibrium value. Both solid phases, of course, must exist together for this to happen.

The actual value of equilibrium p_{O_2} depends on the temperature and if the temperature varies its value also varies accordingly. Similar, type of equilibrium partial pressures of sulphur in the case of sulphides of CO_2 in the case of carbonates, of H_2O in the case of hydroxides, of Cl_2 in the case of chlorides and so on shall be existing and could be evaluated.

In all such equilibria therefore, if equilibrium partial pressure of oxygen (in oxide reaction) and the others in the corresponding reactions, if plotted as it varies with temperature then the equilibrium

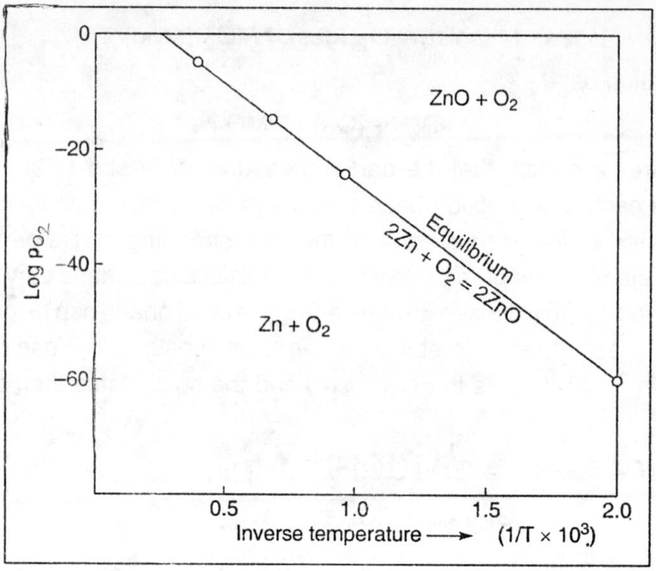

Fig. 14.6 The variation of equilibrium oxygen potential with temperature for a system containing zinc and oxygen and, zinc oxide and oxygen.

gas pressure line shall divide the area in such a way that the compound plus the gas shall exist one side and the metal plus gas shall exist on the other side as shown in Fig. 14.6.

14.4.1 Variation of Free Energy with Temperature

The free energy change in a process is given by

$$\Delta G = \Delta H - T\,\Delta S$$

It means that there may be a change in enthalpy and entropy of the system leading to change in free energy of the system. Although ΔH varies with temperature, the variation being not large, it can be treated as substantially constant over a fairly wide range of temperatures without involving much error. The ΔS however changes with temperature in a process, particularly when gaseous phase is involved along with the condensed phases.

In the generalised chemical process of oxidation of an element as depicted in Eq. (14.14), if the reaction occurs as a natural process, may be after initiation, the enthalpy change of the process shall be negative. Since one g. mole of oxygen gas disappears in this process the entropy must decrease appreciably in ths process.

These changes are qualitatively shown in Fig. 14.7 plotted under standard conditions. The plotting of such a plot needs the knowledge of enthalpy of this reaction and the entropy change accompanying the process. The slope of the $\Delta G°$ vs. T plot is $\Delta S°$. The value of $-T\,\Delta S°$ continuously increases with temperature. If actual examples are taken for calculations and the $\Delta G°$ values for various oxide formation reactions are plotted, they all will show similar straight lines originating from various levels on the free energy axis equivalent to $\Delta H°$ of that reaction. For similar reactions of various oxides the entropy is given by

$$\Delta S° = S°_{(oxide)} - S°_{(metal)} - S°_{(oxygen\ gas)} \qquad ...(15.16)$$

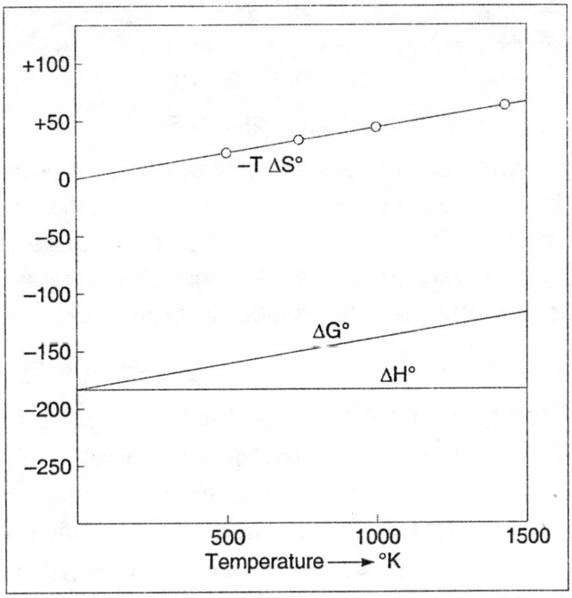

Fig. 14.7 The variation of free energy, enthalpy and entropy for the reaction 4/3 <Cr> + {O_2} = 2/3 <Cr_2O_3> with temperature. ($\Delta G^\circ = -178500 + 41.4\ T$) cal/g mole.

and since $S^\circ_{(oxide)}$ and $S^\circ_{(metal)}$ are practically the same, within the accuracy possible in such calculations, **the entropy change arise predominantly due to disappearance of one g mole of oxygen gas and hence it is negative.** The entropy changes in various such metal oxidation processes are expected to be substantially of the same value. The ΔG° vs. T plots therefore slope upwards, towards less negative values of ΔG°. If the format of the reactions change from that in Eq. (14.14) to something like :

$$2M + O_2 = 2MO \qquad ...(14.19)$$

or

$$4/3\ Al + O_2 = 2/3\ Al_2O_3 \qquad ...(14.20)$$

their slope is likely to change and in that case these lines may even intersect each other.

These are exceptions to this general nature of slopes of these ΔG° vs. T plots in some oxidation reactions. The case of

$$\langle C \rangle + \{O_2\} = \{CO_2\} \qquad ...(14.21)$$

gives entropy change as

$$\Delta S^\circ_{(14.21)} = \Delta S^\circ_{CO_2} - S_C - S_{O_2}$$

Since the gas entropy values are far more than that for the condensed phases and since

$$S_{CO_2} = S_{O_2} \quad \text{and} \quad S_{CO_2} > S_C \quad \text{or} \quad S_{O_2} > S_C$$

and therefore the resultant $\Delta S^\circ = 0$ for the above reaction.

If only means that the ΔG° vs. T plot for the formation of carbon dioxide gas has to be a nearly horizontal line as is the case. This is the exceptional behaviour of $C \rightarrow CO_2$ reaction.

Conversely for a reaction

$$2C + \{O_2\} = 2\{CO\} \quad \text{...(14.22)}$$

$$\Delta S^\circ = S_{CO} - 2S_C - S_{O_2}$$

the result is increase in entropy, as volume is nearly doubled, for the same pressure. The slope of ΔG vs. T shall therefore, be reverse of that of the earlier metal oxidation reaction. This is in fact true. This is yet another exceptional behaviour of $C \rightarrow CO$ reaction and slope being in opposite direction, this line cuts across many other oxide lines and alters the relative stabilities of the oxides *vis-a-vis* the carbon monoxide line and which is fully exploited in extractive metallurgy for metal extraction purposes.

14.4.2 Free Energy Change with Change of Phase

So far the discussion centered around when both the reactant and the product remained in solid state. While probing the free energy temperature relationships the temperature of interest may be high enough such that either the reactant or the product or both melt or evaporate because the temperature under consideration transgresses their melting or boiling points. The question arises as to whether the straight line relationship of free energy and temperature continue to be the same or change at these temperatures when phase changes occur? **In other words will it have the same slope or the slope changes?**

Let us take the case first when the metal or the reactant first melts. The reaction then shall be written as

$$(M) + \{O_2\} = \{MO_2\} \quad \text{...(14.22)}$$

The entropy of the reaction upto the temperature when all are solids has one value as shown in Fig. 14.4. Now the entropy change of this reaction is different from that in Eq. (14.19) as

$$\Delta S^\circ_{(14.22)} = S_{MO_2} - S_M - S_{O_2} \quad \text{...(14.23)}$$

If the Eqs. (14.19) and (14.23) are compared, earlier the entropies of solid meal and solid oxide were equated, but now being different phases, are not that similar to be equated to cancel each other and hence

$$S_{M(liq)} > S_{M(solid)}$$

Hence, ΔS° = more negative than when both were solid,

and in that case if the free energy of Eqs. (14.19) and (14.23) are compared as

all solid phases $\quad \Delta G_1^\circ = \Delta H^\circ - T\,\Delta S_1^\circ$

when reactant is liquid $\quad \Delta G_2^\circ = \Delta H^\circ - T\,\Delta S_2^\circ$

wherein $\quad -T\,\Delta S_1^\circ < -T\,\Delta S_2^\circ$

$$-\Delta G_1^\circ > -\Delta G_2^\circ$$

In other words the pot shall slope further upwards or slope of this ΔG vs. T relationship shall be more as shown in Fig. 14.8 from the point of temperature as melting point.

Since the entropy of melting of many metals has fairly the same value the plots for many metals may again be parallel originating from their respective melting points onwards.

If the temperature of the process of oxidation is such that the reactant boils, the value of $-T\,\Delta S°$ shall be still more negative than that for liquid. In that case the $\Delta G°$ vs T plot shall slope further upwards or more towards less negative value of $\Delta G°$ as shown in Fig. 14.8.

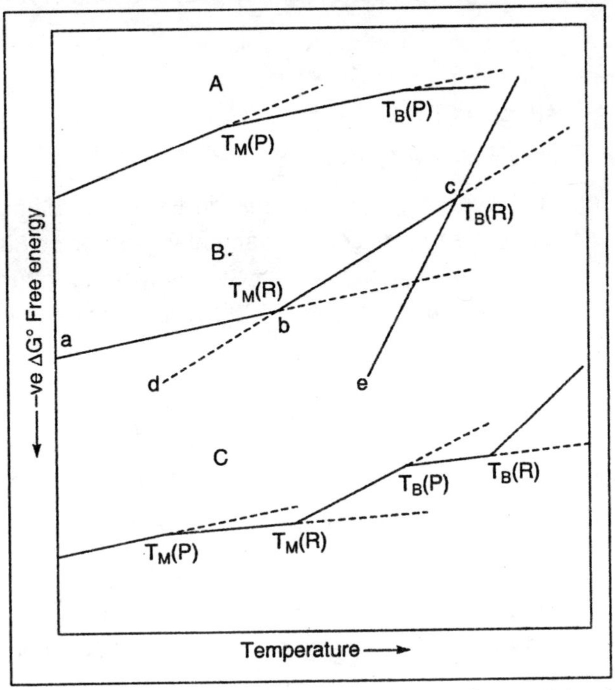

Fig. 14.8 Variation of free energy with temperature as phase changes take place in reactants and products. B is used to denote boiling point and M is used to denote melting point. The subscript R is used for reactant and P is used for product.

Let us now consider that instead of reactant melting, product oxide melts first then the entropy change shall be

$$\Delta S° = S°_{(\text{liquid oxide})} - S°_{(\text{solid metal})} - S°_{(\text{gas})} \qquad \ldots(14.24)$$

= less negative than when both are solids

Therefore, the free energy in this case shall be more negative than when both are solids and as a result the slope of the line shall be more negative than when melting of the reactant occurs. Likewise these lines bend further downwards when it boils. The combination of all these changes have been shown Fig. 14.8 schematically. The actual values for a specific case can be evaluated knowing the entropy values of various phases involved in the reaction.

■ **Example 14.1** For three different forms of reactant metal M, solid, liquid and gas, being oxidised to produce solid metal oxide, three different values of free energy changes were observed as :

(1) 2 <M> + {O_2} = 2 <MO> $\quad \Delta G° = -290400 + 46\,T$

(1) 2 (M) + {O_2} = 2 <MO> $\quad \Delta G° = -358754 + 102\,T$

(1) 2 {M} + {O_2} = 2 <MO> $\quad \Delta G° = -298400 + 55\,T$

How to know which one of these is for which specific state of M?

Solution : The free energy of formation of solid oxide from solid and liquid M should be same at the m.p. In other words the two $\Delta G°$ when equated shall give the temperature at which $\Delta G°$ shall be zero.

Therefore, there are three equation-pairs namely 1-2, 1-3 and 2-3 to obtain the required temperatures.

1-2 – when equated, gives T = 1220°K

1-3 – when equated, gives T = 888°K

2-3 – when equated, gives T = 1305°K

The question still remains, even if 1-3 are solid-liquid transition, being the lowest transition temperature, which one of the 1 or 3 is for solid? It is clear that m.p. must be 888°K, and b.p. 1220°K. The ambiguity can be solved by evaluating values of $\Delta G°$ at any temperature below 888, *i.e.*, say 800°K, in between 888 and 1220°K, *i.e.*, to say that 1200°K and above the b.p. 1220, *i.e.*, say at 1280°K.

The values are as follows :

Temperature	reaction 1	reaction 2	reaction 3
800	–253600	–277154	–254400
1200	–254400	–256754	–243400
1280	–231520	–228194	–288000

All reactions show negative free energy values of $\Delta G°$ and hence can be correlated schematically as shown in Fig. 14.8(b). It is logical to expect that between 1 and 3 at 800°K whatever shows less –ve $\Delta G°$ value shall be liquid as shown by supercooled dotted line. Hence Eq. 1 is for solid and 3 is for liquid. Obviously, 2 is for gas. The proof is that $\Delta G°$ of reaction 2 at 1280°K must be least –ve and which is the case.

The same can be argued in a different way as follows :

The entropy change in each one of these three cases has to be relatively calculated as follows :

(1) When solid M $\quad \Delta S° = S_{(oxide)} - S_{(oxygen\ gas)} - S_{(metal)}$

(2) When liquid M $\quad \Delta S° = S_{(oxide)} - S_{(oxygen\ gas)} - S_{(metal)}$

(3) When gas M $\quad \Delta S° = S_{(oxide)} - S_{(oxygen\ gas)} - S_{(metal)}$

The entropy change therefore shall be in increasing order in the order

$$\text{solid M} \rightarrow \text{liquid M} \rightarrow \text{gas M}$$

The entropy values are given in the three equations as a multiplier of temperature T. Hence the equation of free energy for reaction 1 is for solid, 3 is for liquid and 2 is for gas.

14.5 ELLINGHAM DIAGRAMS

These are the **free energy vs. temperature relationships** which were plotted for various oxidation and sulphidation processes of several metals and gases by H.J.T. Ellingham during the second world war for the first time and these later became known after him as **Ellingham Diagrams.** Now, such diagrams are available not just for oxides and sulphides but for many other compounds like chlorides, carbides, nitrides, carbonates, hydroxides and so on and also for many complex compounds like silicates, sulphates, phossphates, and the like. The Ellingham diagrams for simpler compounds are by far more useful and hence more popular. In each one of this diagram the basic reaction is formation of the relevant compound from respective elements, one of which shall be the anionic part like oxygen, sulphur, chlorine, and so on. These diagrams are known by these anionic part as **Oxide-Elligham-Diagrams, Sulphide-Elligham-Diagrams**, and so on.

Initially only oxide diagrams were prepared but soon the others were also ploted because of their importance. The diagram plotted by Ellingham for oxides was recast by F.D. Richardson immediately after the world war second. It includes not only the ΔG° vs. T relationship but also superimposed on this diagram are the additional axes of p_{O_2}, CO/CO_2 ratio, H_2/H_2O ratio scales and relevant O, C and H points as shown in Figs. 14.1, 14.2 and 14.3. In other words the Fig. 14.1, 14.2 and 14.3 are superimposed on the ΔG° vs. T diagram. The free energy is shown on Y-axis but as per Eq. (14.17) it is also the same as $RT \ln p_{O_2}$. It also means that the present Ellingham diagram for oxides includes the free energy temperature relationships for various metals and provides means to obtain equilibrium CO/CO_2 ratio and equilibrium H_2/H_2O ratio in contact with metal-metal oxide system at any temperature over the range shown in the diagram.

The diagram currently in use and referred universally is shown in Fig. 14.9. The characteristics of this diagram are discussed in detail in the following. These are :

1. Several metal-metal oxide systems of interest in chemical and metallurgical engineering are shown in this diagram.
2. The zero free energy point is at the top and the Y-axis is shown in the form of –ve ΔG° values.
3. This is a standard diagram meaning thereby that the reactants (metals) and the products (oxides) are all in their standard states or pure states, *i.e.*, at unit activities.
4. Almost all oxides have negative free energy of formation. The exceptions being the noble metals like platinum, gold and to some extent silver.
5. As the electrochemical series is scanned from noble to baser metals downwards, nearly similar order, though not exactly matching the same is observed, as the lines are scanned from top to bottom in this diagram. The nobler metals are at the top and slowly change downwards to less and less noble to baser ones towards the lower side.
6. Since the –ve ΔG° moves downwards, the lower is the line the lower is the free energy of that oxide. In other words the lower oxides are relatively more stabler than the ones above it. Obviously the nobler metal oxides, which are most unstable, are at the top and the stabler oxides of the baser metals are at the bottom. This relative stability is true at any given temperature.
7. The carbon-carbon monoxide line on this diagram is exceptional in that it slopes in the direction just reverse to almost all other oxides. In other words the $-\Delta G^\circ$ value for this reaction goes on increasing with temperature. For all other oxides, except carbon dioxide, the $-\Delta G^\circ$ go on decreasing with temperature. The carbon-carbon monoxide line therefore cuts across many of these oxide lines and thereby reverse the relative stability relationships with the corresponding oxide after the intersection.

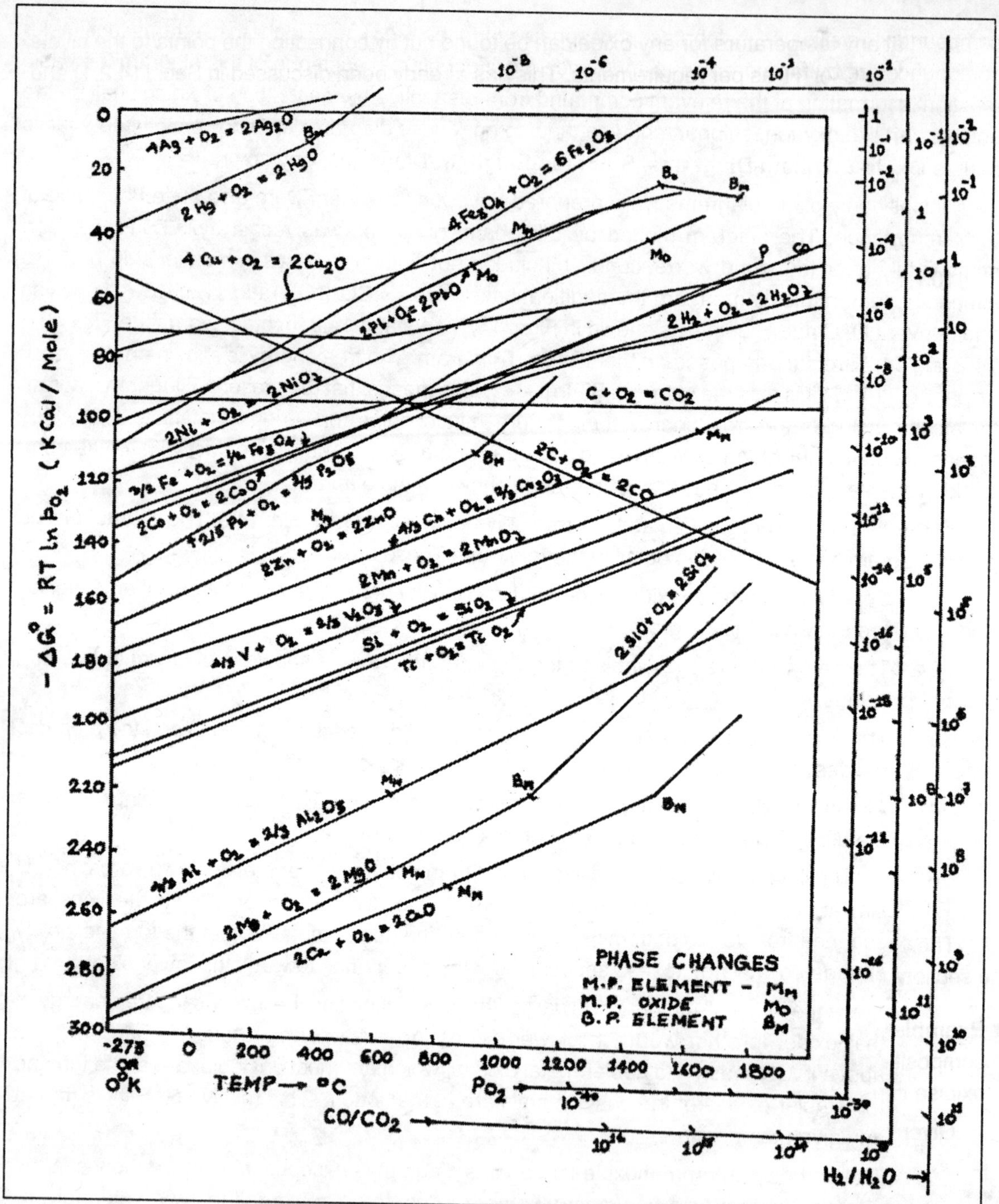

Fig. 14.9 Ellingham diagram for oxide systems.

8. Any point on this diagram means certain temperature and certain value of ΔG° or RT ln p_{O_2}. If this point is joined to either 'O' or 'C' or 'H' points on the graph and extrapolated, can give the equilibrium partial pressure of oxygen or equilibrium ratios of CO/CO_2 or H_2/H_2O in contact with metals and their oxides. Therefore, such equilibrium parameters

at any temperature for any oxide can be found out by connecting the points to the pivots 'O', 'C' or 'H' as per requirements. This has already been discussed in Sec. (14.2.1) and Fig. 14.4.

9. The lines on the diagram for most metals are nearly parallel as discussed in Sec. 14.4.1. The slope of an individual line may alter on melting or boiling of the reactant or product or solid state phase changes therein as discussed in Section 14.4.2. This therefore may result in intersection of some lines and thereby alter their relative stabilities.

10. Free energy is an extensive property and hence quantitative comparison of free energies of formation of oxides can not be made in two different reactions involving different masses. Erroneous conclusions may come out if care is not taken in standardising the two reactions for the sake of comparison. Free energies can be compared based on one g mole of the reactant metal or product oxide or oxygen gas. There is no specific bar on any of these bases or any specific preference on scientific principles. The common thing in all these reactions is oxygen gas. This diagram is widely used to assess the relative possibilities of various oxidation reactions with respect to one-another. Whether one particular element would be in a position to reduce the oxide of a different metal under certain conditions can be decided by writing the two oxidation reactions and subtracting one from the other to obtain the net free energy change from this diagram. In the process of subtraction the oxygen part shall get eliminated, if the base is one g mole of oxygen, straightaway. If this is not the base the two equations will have to be balanced to get the same result. It is therefore easier to have the free energy of reactions plotted per g mole of oxygen. The reactions on the diagram are written such that the reactant is one g mole of oxygen and the equivalent metal and oxide are adjusted accordingly,

as
$$2Mn + O_2 = 2MnO$$
$$Si + O_2 = SiO_2$$
$$4/3\ Al + O_2 = 2/3\ Al_2O_3$$

11. On the whole a huge amount of information is stored in this compact diagram.

The way this information has been exploited or is available for such use shall be dealt with in the sections to follow.

■ **Example 14.2** Steel is made by preferential oxidation of impurities in iron of the nominal-composition C-4%, si-1.5%, Mn-0.7%, P-0.35% and balance iron. If pure oxygen is supplied to oxidise these find out the theoretical sequence of elimination of these impurities.

Given :

(1) $2\langle C\rangle + \{O_2\} = 2\{CO\}$ $\Delta G° = -58800 - 31.9\ T$

(2) $[Si] + \{O_2\} = \langle SiO_2\rangle$ $\Delta G° = -217000 - 48.8\ T$

(3) $2[Mn] + \{O_2\} = \langle MnO\rangle$ $\Delta G° = -190800 - 39.25\ T$

(4) $4/5[P] + \{O_2\} = 2/5\langle P_2O_5\rangle$ $\Delta G° = -93100 - 64.5\ T$

Solution : The free energy changes at 1600°C, the usual steelmaking temperature, are found to be

$$\Delta G^{o}_{C} = -118500 \text{ cal/g. mole}$$

$$\Delta G^{o}_{Si} = -125600 \text{ cal/g. mole}$$

$$\Delta G^{\circ}_{Mn} = -\ 117300 \text{ cal/g. mole}$$

$$\Delta G^{\circ}_{P} = +\ 27700 \text{ cal/g. mole}$$

Assuming the product oxide activities to be unity in all the cases and that the oxygen activity shall be the same in all the cases that element shall oxidise first whose equilibrium parallel pressure shall be the minimum. It also means that the sequence of oxidation shall be in the order first that element whose free energy change on oxidation is the minimum. Therefore, the strict sequence shall be first silicon, then carbon, then manganese and lastly phosphorus. In fact the first three are so close that they may oxidise simultaneously and for phosphorus oxidation to be possible the activity of the oxide will have to be made more negative to make it possible.

14.5.1 Thermal Dissociation of Oxides

It has already been noted that in a standard Ellingham diagram the oxide lines slopes upwards from their pivots at $\Delta G^{\circ} = \Delta H^{\circ}$ at zero degrees absolute. It means that as temperature rises the ΔG values become less and less negative. A situation can arise when the line intersect the horizontal line at $\Delta G^{\circ} = 0$, *i.e.*, $p_{O_2} = 1$ atmos at some temperature. When it does so it means that the metal and metal oxide are in equilibrium at one atmospheric pressure of oxygen and the $\Delta G^{\circ} = 0$. In reality in a system open to atmospheric air wherein $p_{O_2} = 0.21$ the non-standard $\Delta G = 0$ position may be reached much earlier. Any further rise in temperature would make the ΔG° now +ve, meaning thereby that oxidation of metal is no longer feasible. On the contrary oxide shall dissociate into metal and oxygen. In fact this is the ideal process of extraction of metal from its oxide merely by heating, provided it is possible to separately collect the metal so reduced. The process can be typically illustrated by the example of silver oxide.

For $\quad 4 \langle Ag \rangle + \{O_2\} = 2 \langle Ag_2O \rangle \qquad \Delta G^{\circ} = -7300 + 15.76\ T$ ...(14.25)

and $\quad \Delta G^{\circ} = 0 \quad \text{at} \quad T = 426°K$

Hence, above 462°K silver oxide shall dissociate into silver and oxygen. This process can be used for extraction of silver from its oxide merely by thermal dissociation. Unfortunately silver is not available as oxide, at least not in natural form, to make it a commercial possibility.

Theoretically, all these lines which originate from below point O in the oxide Ellingham diagram shall intersect the $\Delta G^{\circ} = 0$ line at some temperature thus thermally dissociate at above that temperature where it intersects the $p_{O_2} = 1$ line. In normal atmospheric air the situation shall change accordingly. Normally, these temperatures are fairly large and even if the oxide dissociates at such high temperatures it will not be possible to collect it in pure metallic form because of its highly reactive nature at that high temperature. The end result may therefore, be that the oxide does dissociate into metal and oxygen at one moment and immediately there after, while trying to separate the metal, it gets oxidised and produces oxide again.

Similar considerations are possible for other diagrams as well. Mercury occurs in nature as Hg_2S (mineral cinnabar) and has only small –ve value of ΔG° of formation. If this is heated around 500°C it does dissociate into Hg and S, both gaseous forms, and can be condensed to recover Hg. This is used in commercial practice to win mercury from its sulphide ores.

14.5.2 Intersecting Lines on Ellingham Diagram

Althogh by and large the lines on the oxide diagram below the melting point of both the metal or its oxide are parallel to each other, their slope undergoes changes at the melting or the boiling points of the metal or its oxide. Since the melting and the boiling points for different metals are quite different, the change in slope in these lines takes place at different temperatures. The slopes also change when phase transformations in solid state occur. The net result can be that some lines may intersect each other at some temperatures. The intersection changes the relative positioning of these lines and thereby its relative stability. It means that $\Delta G°$ values relatively change their order on changes in the phases of the reactants and the products. The situation is described qualitatively in Fig. 14.10.

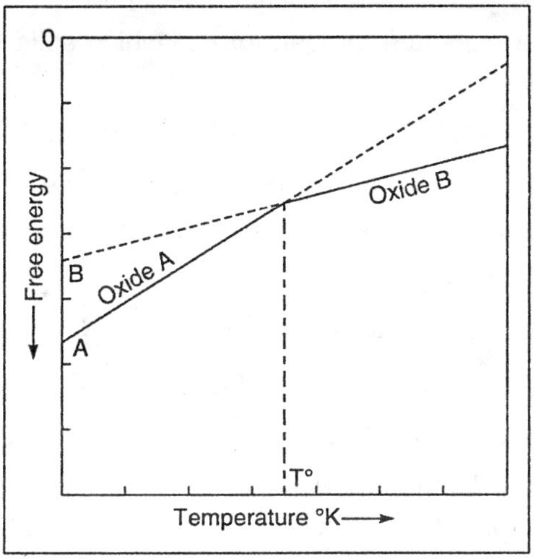

Fig. 14.10 Intersecting lines on the Ellingham diagram indicating the reversal of relative stabilities after the temperature of intersection.

Let us assume that the two lines of oxidation of metal A and B intersect at temperature T°K as shown. Therefore, below the temperature 'T' A-oxide is more stable, being lower, than B-oxide. But after the intersection, *i.e.*, above temperature 'T' the situation is just the reverse. Hence below 'T' reaction of reduction of A oxide by B is not possible whereas above temperature 'T' it is readily feasible. Such situations can be gainfully exploited to carry out certain reactions which otherwise are not possible below a certain temperature.

This is possible because, free energy consists of two parts,one enthalpy and the other entropy; enthalpy being substantially constant with temperature but entropy changing with temperature. The situation becomes so favourable after a certain temperature that the net result in these two different cases is reverse. This will be discussed further again later.

14.5.3 Carbothermic Reduction of Oxides

The effect of intersection of lines on the oxide Ellingham diagram has just been discussed. The C-CO line is such that it cuts across lines of many oxides of interest. The slope of C-CO line is reverse to those of other oxides and therefore it should be possible to reduce many oxides by carbon above the temperature at which the C-CO line intersects their oxide lines. It is thus possible to reduce the oxides of copper, lead, iron, nickle, cobalt, molybdenum, tin, phosphorous, zinc, chromium, manganese, vanadium, tungsten, silicon, magnesium, aluminium, calcium and so on at increasing temperatures in that order using carbon as a reducing agent. Carbon being very cheap, the form of either charcoal or coke is used in fully exploiting this technique for commercial production of these metals from their oxides.

Metals like aluminium or titanium can not be extracted by carbon reduction because these metals react with carbon and the product obtained is in their carbide rather than the metal form. In some cases the temperature required for reduction is so high that it is not a commercial proposition against other cheaper alternatives.

These reduction processes have to be carried out at high temperatures and hence these are known as carbothermic reduction process.

■ **Example 14.3** Find out the minimum temperature at which solid silica would be reduced by carbon to produce liquid silicon.

Given : $2 <C> + \{O_2\} = 2\{CO\} \quad \Delta G° = -53400 - 41.9\,T$ cal/g. mole

$<Si> = (Si) \quad \Delta G° = 10050\,T$ cal/g. mole

$<Si> + \{O_2\} = <SiO_2> \quad \Delta G° = -208000 + 43.3\,T$ cal/g. mole

Solution : From Hess's law for the reaction

$$<SiO> + 2<C> = (Si) + 2(CO)$$

$$\Delta G = +208000 - 43.3\,T + 10050 - 53400 - 41.9\,T$$

$$= +162600 - 85.2\,T$$

This reaction shall be in equilibrium when $\Delta G° = 0$ and therefore this will happen at T = 1635°C which is the minimum temperature at which carbon will be able to reduce solid silica.

14.5.4 Metallothermic Reduction of Oxides

The fact the oxide whose line is below is more stabler than that whose line is relatively above can be utilised in devising processes to reduce oxides that are above by the metal whose oxide is below.

■ **Example 14.4** Thus if $2<Mn> + \{O_2\} = 2<MnO>$ ...(14.26)

$$\Delta G^{o}_{(15.20)} = -184000 + 34.63\,T$$

and $<Si> + \{O_2\} = <SiO_2>$ ...(14.27)

$$\Delta G^{o}_{(14.27)} = -208000 + 43.3\,T$$

Hence the free energy for reaction

$2<MnO> + <Si> = 2<MnO> + <Si>$ ...(14.28)

is obtained by subtracting one from the other as

$$\Delta G^{o}_{(14.28)} = -24000 + 8.67\,T$$

or $\Delta G^{o}_{1600°K} = -10128$ cal/mole

Since this is negative this reaction is feasible or manganese oxide shall be reduced by silicon.

Similarly, commercially uranium oxide is reduced by calcium to produce uranium. There are many such commercial processes. The thermit process, wherein iron oxide is reduced by aluminium to produce molten iron, which is used as a filler metal for welding large objects. Ferro-alloys which are nothing but iron and some other metal combination like, ferro-manganese, ferro-chrome, ferro-silicon, ferro-aluminium, ferro-titanium, ferro-tungsten, ferro-molybdenum, ferro-nickle and host of such alloys which are produced by using reducers like Si, Al, Ca, etc., individually or in combination, to reduce their oxides. The reducers so used have fairly negative ΔG° values of oxide formation as compared with the values of the oxides to be reduced. In other words elements at the lower side of the diagram are used as reducers to reduce the oxides which are relatively above.

■ **Example 14.5** Example given in Section 2.10.2 may be recalled here. It was for calculation of enthalpy. In place of ΔH if ΔG values are inserted it could show the method of calculation of metallothermic reduction of oxides.

14.6 ELLINGHAM DIAGRAMS FOR NON-STANDARD CONDITIONS

The pattern of lines on the standard Ellingham diagram is well established since the reactants and the products are assumed to be in their standard states, *i.e.*, at unit activities. What will happen if the activities are less than one, *i.e.*, the reactants and the products, individually or both are under non-standard conditions?

The free energy change for the reaction in Eq. (14.14) is given as

$$\Delta G = \Delta G^\circ + RT \ln \frac{a_{MO_2}}{a_M \times p_{O_2}}$$

If all the reactants and the products are in their standard states then the RT ln term becomes zero and

$$\Delta G = \Delta G^\circ$$

But if M is not in standard state then

$$\Delta G = \Delta G^\circ + RT \ln \frac{1}{a_M \times p_{O_2}}$$

The values of the product $\dfrac{1}{a_M \times p_{O_2}}$ depends only on temperature because it is nothing but equilibrium constant and whose value is fixed for a fixed temperature. In order to maintain the value of K, *i.e.*, the above quotient the same even if activity of reactant decreases then the equilibrium partial pressure of oxygen shall proportionately increase. By the same logic if the reactant activity is unity but that of product oxide decreases then the p_{O_2} value would decrease to maintain the value of the quotient the same for a fixed temperature. In other words the equilibrium partial pressure of oxygen shall decrease if the product activity decreases.

The situation is shown in Fig. 14.11 by plotting the standard ΔG° vs. T line for a generalised reaction as in Eq. (14.14). The temperature T is marked on this line by point 'A'. This point is connected to point O, which when extended upto the p_{O_2} axis shows the corresponding p_{O_2} value as depicted by point 'A'. Because of decrease in the activity of reactant (metal) the equilibrium p_{O_2} increases to say 'B' value. The straight line joining B with O has a free energy value equivalent o point B at the same temperature T. If this exercise is repeated for several temperatures the ΔG for this non-standard oxidation reaction as it varies with temperature shall be available. The enthalpy

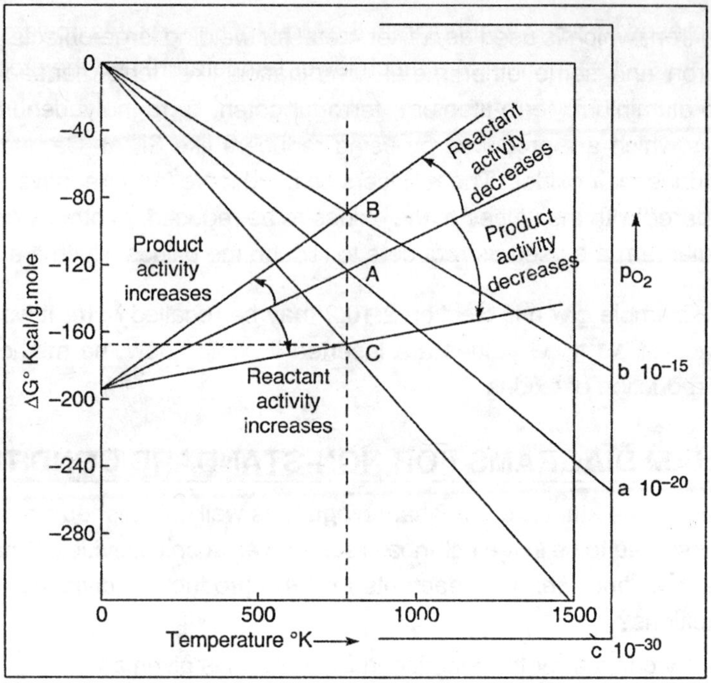

Fig. 14.11 The rotation of lines on Ellingham diagram when the activities of products and/or the reactants change.

change even in non-standard conditions remains the same since it is independent of activities. This non-standard line is thus shown originating at the same point, *i.e.*, at $\Delta G° = \Delta H°$ and is shown as one passing through the point B. By the same logic when the activity of product oxide decreases keeping the metal activity at unit the line originating $\Delta G° = \Delta H°$ and passing through point C shall be obtained.

In fact these findings can be qualitatively stated as: **'If the activity of the reactant decreases the standard free-energy-temperature plot shall rotate anti-clockwise and if the product activity decreases it shall rotate in clockwise direction from the pivot, $\Delta G° = \Delta H°$ on the free energy axis'.**

If both, the activities of the product as well as the reactant decrease then the resultant of their effect shall be the final state of affairs. Coincidentally if may so happen that their variation cancel each other and the standard line may remain the same even in the non-standard conditions

14.6.1 Specific Non-Standard Reactions

The rotation of the standard line o the Ellingham diagram because of non-standard conditions of the reactant and the product may lead to drastic changes in the relative positioning of the various lines. If the relative positions of the lines undergo reversals, the processes otherwise not possible, may become possible as was shown in Sec. 14.3.2 for intersecting lines. Several such processes have been developed and carried out in practice in non-standard conditions, which otherwise are not possible under the standard conditions. Some such specific cases are dealt with below.

It is a fact that solid MgO is far more stable than SiO_2 and hence solid MgO cannot be reduced by silicon at any temperature. But if the activity of SiO_2 is decreased by combining it with CaO the standard line of Si-SiO_2 rotates downwards considerably. At the same time if the activity of magnesium is decreased by using vacuum then the Mg-MgO line rotates upwards. The net result is that the two lines intersect each other at a much lower temperature and the process :

$$\underset{\text{Dolomite}}{CaO.MgO} + \underset{\text{Ferro silicon}}{Fe.Si} = \underset{\text{Slag silicate}}{CaO\text{-}SiO_2\text{-}FeO} + \underset{\text{Vapour}}{Mg} \text{ (Vacuum)}$$

is readily possible and is carried out under vacuum at around 1200°C commercially to produce magnesium. This is known as Pigeon's Process. The relevant thermodynamic data are :

$$2\{Mg\} + \{O_2\} = 2\,(MgO) \qquad (14.29)$$

$$\Delta G° = -358700 + 102.67 \text{ cal/g mole}$$

$$<Si> + \{O_2\} = SiO_2 \qquad ...(14.30)$$

$$\Delta G° = -208300 + 43.3\,T \text{ cal/g mole}$$

Therefore, the reaction

$$2<MgO> + <Si> = <SiO_2> + 2\{Mg\}_{(1\ atmos)} \qquad ...(14.31)$$

is obtained from combination of (14.29) and (14.30). If the reaction is to be carried out under vacuum of 10^{-3} atmospheric pressure then (14.31) will reduce to

$$2<MgO> + <Si> = <SiO_2> + 2\{Mg\}_{10^{-3}\ atmos} \qquad ...(14.32)$$

The $\quad \Delta G^{o}_{(14.32)} = \Delta G^{o}_{(14.30)} - \Delta G^{o}_{(14.29)}$

and $\quad \Delta G_{(14.32)} = \Delta G^{o}_{(14.30)} - \Delta G^{o}_{(14.29)} + RT\ln p_{Mg} \qquad ...(14.33)$

Putting the actual values it becomes

$$= +130400 - 59.3\,T + RT\ln 10^{-3}$$

$$= +130400 - 59.3\,T + 4.575 \times T \times (-3)$$

$$= +130400 - 59.3\,T - 13.72\,T$$

If CaO is added to the reactant the resultant reaction would be

$$2<MgO> + 2<CaO> + <Si> = <2CaO.SiO_2> + 2\{Mg\}_{10^{-3}\ atmos} \qquad ...(14.34)$$

and $\quad 2<CaO> + <SiO_2> = <2CaO.SiO_2> \qquad ...(14.35)$

therefore, $\quad \Delta G^{o}_{(14.35)} = -30200 - 1.2\,T$

and hence $\quad \Delta G_{(14.34)} = \Delta G_{(14.32)} + \Delta G_{(14.35)}$

$$= +100200 - 71.82\,T$$

At equilibrium $\quad \Delta G_{(14.34)} = 0$

therefore, the temperature at which equilibrium holds true is

$$+1100200 - 171.82\,T = 0$$

or $\quad T = 1395°K = 1122°C$

No wonder in commercial practice the extraction of Mg is carried out around 1200°C under a vacuum of 10^{-3} atmos. This is shown in Fig. 14.12.

The other typical example is reduction of silica by high manganese ferro-manganese. The silicon line is well below manganese line in a standard Elligham diagram and therefore reduction of solid silica by manganese under standard conditions does not arise. However, if silicon is dissolved

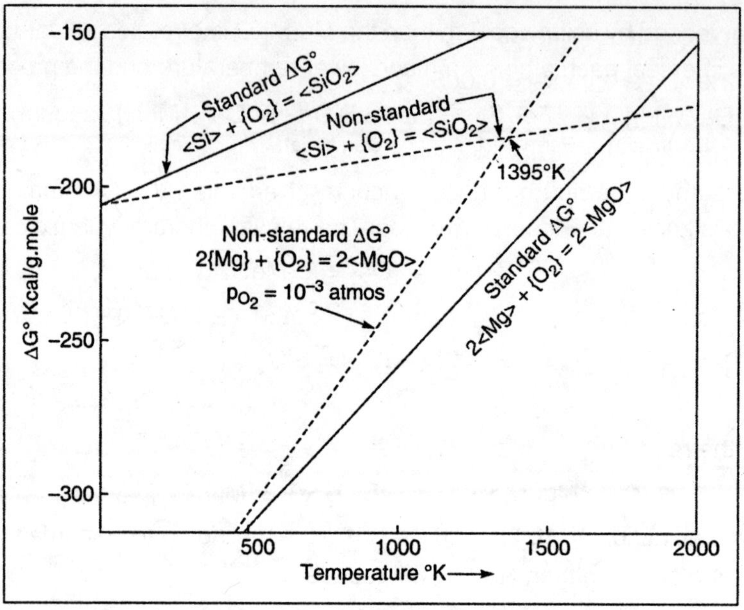

Fig. 14.12 Free energy considerations in the Pidgeon process for the extraction of magnesium by reduction of magnesia by silicon.

in iron the activity of silicon is reduced drastically and hence the silicon-silica line shall rotate upwards. Concurrently as the MnO activity is reduced because of its combination with silica as manganese silicate the Mn-MnO line shall rotate downwards. The net result is that the two intersect at reasonably low temperature like 100°K and hence it is possible for the reaction

$$[Mn]_{Fe\text{-}Mn} + [SiO_2]_{sand\ mould} = [Si]_{Fe\text{-}Mn} + (MnO\text{-}SiO_2)_{slag}$$

to occur in real practice. For the reaction

$$2\,[Mn] + \{O_2\} = 2(MnO) \qquad \text{...(14.36)}$$

$$\Delta G^\circ_{(14.36)} = -\,162200 + 25.25\,T$$

and

$$[Si] + \{O_2\} = \{SiO_2\} \qquad \text{...(14.37)}$$

$$\Delta G^\circ_{(14.37)} = -\,215400 + 47.73\,T$$

hence for

$$2\,[Mn] + (SiO_2) = [Si] + 2\,(MnO) \qquad \text{...(14.38)}$$

$$\Delta G^\circ_{(14.38)} = \Delta G^\circ_{(14.36)} - \Delta G^\circ_{(14.37)}$$

$$= +\,53200 - 22.48\,T$$

The equilibrium constant of reaction (14.38) is

$$K_{(14.39)} = \frac{a_{[Si]} \times a^2_{(MnO)}}{a^2_{[Mn]} \times a_{(SiO_2)}}$$

$$= \frac{[\%Si]\,.\,f_{Si} \times N^2_{MnO} \cdot \gamma^2_{MnO}}{[\%Mn]^2\,.\,f^2_{Mn} \times N_{(SiO_2)} \cdot \gamma_{SiO_2}}$$

Assuming approximate value as

(i) % Si = 2% $f_{Si} = 0.005$

(ii) % Mn = 75% $f_{Mn} = 0.01$

(iii) $N_{MnO} = 0.1$ $\gamma_{MnO} = 0.5$

(iv) $N_{SiO_2} = 0.9$ $\gamma_{SiO_2} = 0.8$

Putting these value in equation (14.38) gives

$$K_{(14.38)} = \text{approximately } 10^{-5}$$

$$\Delta G_{(14.38)} = \Delta G^{\circ}_{(14.38)} + RT \ln K$$

$$\Delta G_{(14.38)}\ (1600°K) = 16800 + 4.575 \times 1600 \times \log 10^{-5}$$

approximately $= -20000$ cal

Since free energy change is fairly –ve the reaction of reduction of silica by Mn is possible and which is shown in Fig. 14.13.

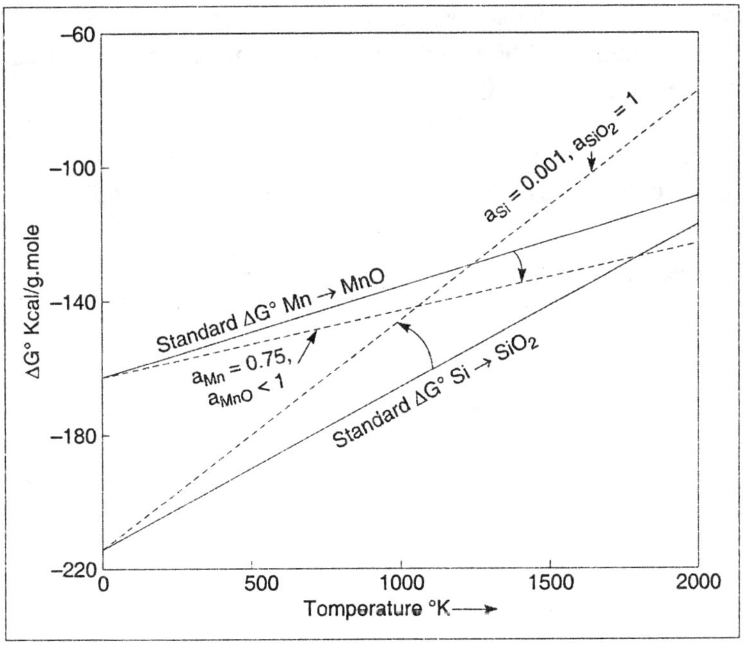

Fig. 14.13 The free energy-temperature relationships indicating the possibility of reduction of silica by manganese.

14.6.2 Reactions Under Vacuum

This belongs to the category of reactions under non-standard conditions but being of its own type has been described separately. Pressure is also one of the standard state and 1 atmos. is taken as the standard state for plotting the Ellingham diagram. If the reactions are carried out at less than 1 atmos, *i.e.*, under vacuum or at higher than 1 atmos pressure the standard lines, since all of these reactions involve a gaseous component, *i.e.*, at least oxygen, undergo changes with pressure. For a general reaction as shown in Eq. (14.14), if pressure is reduced, the oxide would tend to get reduced and produce metal and vice versa. But when the metal and its oxide are not at unit activities and in that case if the oxygen partial pressure is decreased then the equilibrium shall move in such a way as to maintain the equilibrium constant the same and vice-versa.

Pressure is important when gaseous product is formed in these reactions and which is possible only when carbon-monoxide or carbon-dioxide are formed. Carbon dioxide is not a high temperature stable phase and hence only monoxide is considered here. The decrease in the partial pressure means decreased p_{CO} and hence for the same activity of carbon the equilibrium oxygen partial pressure has to decrease or the C-CO line shall rotate in clockwise direction. It means that this line shall cut the oxide lines at still lower temperatures and thereby make the carbothermic reduction possible at lower temperatures. If the pressure is increased above one atmosphere then just the reverse shall happen. It also means that at lower pressures, *i.e.*, under vacuum carbon will be readily oxidised by oxygen and that the product of activities of oxygen and carbon shall be decreased. This is fully exploited in carrying out either decarburisation of steel melt under vacuum by supplying oxygen or deoxidation of steel bath by supplying carbon readily. This is what is known as vacuum treatment of molten steel. This pressure dependency of carbon oxidation reaction can pose serious threat because of the possibility of disintegration of refractions (oxides) by carbothermic reduction under vacuum.

■ **Example 14.5** Find out the levels of carbon that can be achieved when molten steel is decarburised by supply of oxygen at 1 atmosphere pressure and under vacuum of 10^{-3} atmosphere.

Given : The oxidation of carbon in steel melt takes place as

$$[C] + [O] = \{CO\}$$

and

$$\log K = \frac{1056}{T} + 2.131$$

Solution : The value of equilibrium constant at the steelmaking temperature or 1600°C or 1873°K is given as

$$\log K_{1873°K} = 2.69 \quad \text{or} \quad K_{1873°K} = 2.02 \times 10^{-3} \text{ atmos.}$$

K can also be expressed for the above reaction as

$$K = \frac{p_{CO}}{a_{[C]} \times \alpha_{(O)}}$$

The level of oxygen in both the cases shall be proportional to the oxygen pressure supplied and it is the same in either case and therefore, the carbon level achievable shall be proportional to the pressure of carbon monoxide in the surrounding atmosphere. Therefore, assuming that Henry's law is obeyed then

$$K_{(\text{ atmos})} = \frac{1}{[\%C] \times [\%O]} = 2.02 \times 110$$

therefore, %C in steel melt at under one atmos will be about 0.3-0.4%

and %C in steel melt at 10^{-3} atmos will be about 0.1.

This shows any vacuum treatment is given to steel to reduce carbon to a very low level, which otherwise is not achievable in open atmosphere.

14.7 CONCLUDING REMARKS

The free energy temperature relationships in the form of Ellingham diagrams is a fundamental knowledge in the understanding of oxidation reduction reactions for predicting the equilibria and which in turn is essential to understand the smelting and refining processes.

□□□

Chapter 15

Kinetics of Processes

Rates of Reactions

15.1 INTRODUCTION

So far the reader is quite conversant with state and path properties. Thermodynamics mainly deals with state properties and the conclusions therefrom are independent of path of the process. Taking the example of overhead water tank, one can say that the energy exchange associated with the transfer of water from tank to the ground level is fixed no matter whether water flows down, via 15 mm dia pipe or 100 mm dia pipe or through a narrow capillary like tube (disregarding the frictional component in each). Whereas in each of these cases the time required for flow of a certain amount of water depends on the path, *i.e.*, the diameter of the tube carrying the water. The rate of flow shall further get complicated if the section of the tube was non-uniform and having various restrictions in the form of bends and the like. The rate of flow of water therefore a path function. The study of rate of any such thermodynamic change constitutes the subject matter of the field of '**kinetics**'.

The study of rates of chemical or physicochemical processes ought to be very rigorous and in detail in order to appreciate the processes and evaluate the time taken for a certain change to occur. Conversely during a certain time interval it would be possible to evaluate the extent of change that might occur, from such kinetic studies. The industrial or commercial reactors wherein the processes are carried out in actual practice, are often fairly complicated and quite often these are studied in simplified models in laboratory and the findings are compared with the commercial results to assess the closeness of such simulations.

The possibilities of a process occurring in practice is indicated by its –ve free energy change or decrease in the free energy. In reality a thermodynamically feasible process may occur so slowly that apparently it might give an impression of not occurring of all. The fact, in such cases is, that the process is taking place at so low a rate that within the reasonable time no perceptible change is observed. This is best illustrated from the example of a mixture of $H_2 + O_2$ contained in a vessel. Factually H_2 and O_2 are reacting but the rate of such reaction is negligibly small.

Kinetic, studies may be performed under '**steady**' or '**unsteady**' status of the system. At this stage the discussion is confined to **steady state condition** only. Processes if occur within a single homogeneous phase, the study is referred to as study of **kinetics of homogeneous reactions** but if it is between two or more phases then it is the case of **kinetics of heterogeneous reaction.**

The homogeneous kinetics is quite uncommon whereas heterogeneous kinetics is normally possible. The examples of systems like solid-solid, solid-gas, liquid-liquid, etc. are all examples of heterogeneous systems. Sometimes a new phase is formed during the process and hence the role of nucleation and growth may dominate the kinetic studies in such cases. The formation of CO bubbles in steelmaking, precipitation reactions, etc. are typically the best illustrations of this category.

Since kinetics is path dependent, the process may be sub-divided into various sub-steps to make up the overall process. These are called elementary or kinetic steps. This is referred to as study of mechanism of reaction and rate of the change in each of these sub-step may not be the same depending upon the nature of these sub-steps.

In such a case the overall rate of the process will be dictated by the rate of the slowest kinetic step in the overall path. This is known as the rate controlling step and which is revealed through the study of mechanism of the process. In a given overall mechanism the rate of the rate-controlling step is virtually the maximum rate the process shall exhibit approximately. **The aim of the kinetic studies is therefore to know the overall rate of the process under a given set of conditions (*i.e.*, path), to evaluate its mechanism and therefrom the rate controlling step and, to predict the condition under which the rate controlling step would be enhanced to achieve higher overall rate of the process.**

The rate or velocity of a reaction is defined as the rate of change of concentration (decrease generally) of reactants and (generally increase) or products of the reaction. If 'C' is the concentration of the reactants then the rate is given by $-dC/dt$. The reaction rates of some reactions like the thermit reaction may be extremely fast or for others may be quite slow. It is a common observation that the rate is a function of concentration and temperature.

15.2 KINETICS OF HOMOGENEOUS REACTIONS

The reactions which take place within a single phase are known as homogeneous reactions. Solid state homogeneous reactions are just not possible. Liquid-state homogeneous reactions may occur but rather uncommon. However gaseous homogeneous reactions are not only possible but numerous, *e.g.*,

$$2H_2 + O_2 = 2H_2O \text{ at some temperature above } 100°C$$

or

$$H_2 + Cl_2 = 2HCl \quad \text{and so on.}$$

These are often catalytic reactions meaning thereby that catalyst help the reactions to occur readily, *i.e.*, accelerate them immensely.

15.2.1 Order of Reaction

The sum of the powers to which the concentration of the reacting atoms or molecules that have to be raised to determine, the rate of the reaction is known as the **'order of reaction'**. The order of a reaction is quite independent from molecularity of the reaction. If the process is expressed as

$$aA + bB + \rightarrow \text{product} \qquad ...(15.1)$$

then the rate of the reaction is expressed as

$$-\frac{dC}{dt} = k \times C_A^a \times C_B^b \cdots \qquad ...(15.2)$$

where k is the **'velocity constant or specific reaction rate'**, or **'rate constant'** and C_A^a, C_B^b, ... are concentrations at time t. The **value of (a + b + ...) is then the order of the reaction.**

For a first order reaction

$$-\frac{dC}{dt} = k \times C_A \times C_B \cdots \qquad ...(15.3)$$

and since the concentrations of A or B or ... are going to vary in proportion to one another

$$-\frac{dC}{dt} = k \times C_A \qquad ...(15.4)$$

or

$$-\int \frac{dC}{C_A} = k \int dt \qquad ...(15.5)$$

On integration within limits $C = C_o$ at $t = t_o$ and $C_t = C_t$ at $t = t$

$$k = \frac{2.303}{t} \log \frac{C_o}{C_o - r} \qquad ...(15.6)$$

where r is that amount of C reacted in time t. Similarly, for a reaction if

$$-\frac{dC_A}{dt} = -\frac{dC_B}{dt} = k \times C_A \times C_B \qquad ...(15.7)$$

shall be a second order reaction. In this way the higher order reaction can be evaluated. The order can be zero also if the rate does not depend on concentration of the reactants. Also it can be a fraction depending upon the reaction.

For a **zero** order therefore dr/dt = k or k = r/t and a straight line relationship exists between r and t. For a half order reaction $dr/dt = k(C_o - r)^{1/2}$ and $k = 2/t\,[C_o^{1/2} - (C_o - r)^{1/2}]$ and hence t vs. $[C_o^{1/2} - (C_o - r)^{1/2}]$ should be a straight line. Likewise for first order reaction $dr/dt = k(C_o - r)$ or $k = 1/t\,[\log (C_o/C_o - r)]$ and t vs. $[\log (C_o/C_o - r)]$ should be a straight line relationship. For one order reaction $dr/dt = k(C_o - r)^2$ or $k = 1/t\,[r/C_o(C_o - r)]$ and t vs. $[r/C_o(C_o - r)]$ has to be a straight line relationship.

The order can thus be verified from experiment data by plotting the respective variations as a function of time. For example if log $[\log (C_o/C_o - r)]$ vs. time is a straight line it is first order process and so on.

■ **Example 15.1** The rate constants for the leaching of alumina in aqueous NaOH solution at different temperatures are :

log k	−12.9	−13.1	−13.3	−13.5	−13.7	−13.9
$1/T \times 10^3$	2.25	2.30	2.35	2.40	2.45	2.50

where k is the rate constant and T is in 0°K. Calculate the activation energy of the process.

Solution : According to Arrhenius equation on log scale

$$\log k = \log A - \frac{0.4342}{T} \times \frac{Q}{T}$$

where the factor 0.4342 converts the natural log to log to the base ten. The slope is therefore equal to

$$\frac{-13.1+12.9}{2.30-2.25}\times 10^3 = -4\times 10^3$$

This is also equal to $\frac{0.4342}{R}\times Q$ from the equation and hence

$$\frac{0.4342}{R}Q = -4\times 10^3 \text{ or } Q = -1800 \text{ cal/g mole.}$$

The –ve sign indicates only the –ve nature of the slope in this exercise. The data are such that the same value is true even if other pairs are taken into account because the relationship is exact straight line.

■ **Example 15.2** For kinetic studies of a certain process following data were obtained at certain temperature.

Time min.	2	4	8	12	15	20	25	35
%wt loss (cumulative)	15	27	38	52	63	77	92	99

(the %wt loss shown above is of reactants). Find out the order of this reaction and also find out the time required to achieve 50% of the reaction.

Solution : The data are analysed further by evaluating the following values :

Time min.	2	4	8	12	15	20	25	35
%wt loss (cumulative)	15	27	38	52	63	77	92	99
%(100 – r) (unreacted)	85	73	62	48	37	23	8	1
$[C_o^{1/2}-(C_o-r)^{1/2}]$	0.79	1.46	2.13	3.08	3.92	4.81	7.18	9
$\log C_o/(C_o-r)$	0.070	0.136	0.207	0.318	0.431	0.638	1.079	2
$\frac{r}{C_o}(C_o-r)$	12.75	19.71	23.56	24.96	23.31	17.71	7.36	1.98
$\frac{2}{t}[C_o^{1/2}-(C_o-r)^{1/2}]$	0.79	0.58	0.53	0.51	0.53	0.48	0.57	0.51

the average value of $\frac{2}{t}[C_o^{1/2}-(C_o-r)^{1/2}] = 0.562$. These data are developed to plot graphs of time t vs. each one of the values given by horizontal columns as shown in Fig. 15.1. Referring back the conditions for straight line relationships for certain order of a reaction now gives that since in the graph $[C_o^{1/2}-(C_o-r)^{1/2}]$ vs time t is almost a straight line relationship whereas the other relationships are far away from this nature therefore the process must be an *'half order'* reaction. Since for such reaction the rate constant $k = \frac{2}{t}[C_o^{1/2}-(C_o-r)^{1/2}]$ and since these values vary to some extent the average value can be taken as 0.562.

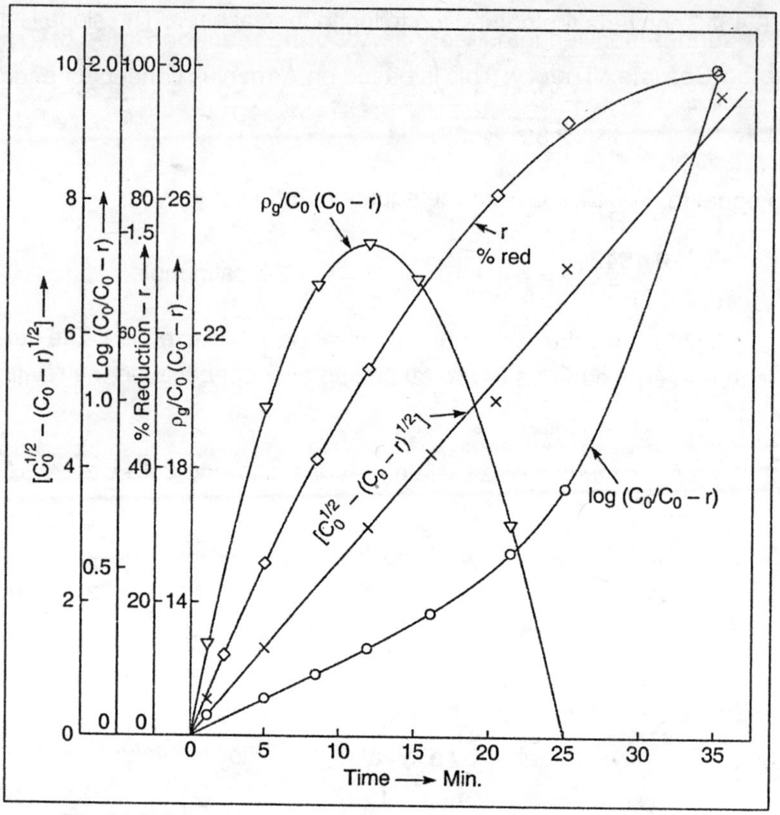

Fig. 15.1 Quantitative method for evaluation of order of a reaction.

The time required for 50% of the reaction to occur can be calculated by putting the values in the above equation wherein all values except that of time 't' are known as k is 0.562 and 'r' is 50% and C_o is 100% and hence

time t for 50% reduction = 10.42 min.

15.3 THEORIES OF REACTION RATES

The earlier collision theory of reaction rates is based on the assumption that reaction occurs because of collision of a reactant atoms or molecules with a certain minimum intensity of impact. The minimum impact energy is nothing but activation energy of the reacting atoms or molecules and it depends on the concentration (mass) and temperature of the reactants. For collision to be successful in making the reaction take place the orientation of the colliding atoms or molecules should be appropriate. Hence the rate

$$\text{Rate} = O_P \, . \, N_c \, . \, e^{-Q/RT} \text{ atoms/moles/cc/sec} \qquad ...(15.9)$$

where N_c is the collision per sec and Q is the activation energy and O_P is the orientation probability or steric factor. This equation has the same form as that of Arrhenius equation where $O_P \times N_c$ is constant, indicating the probability of a reacting particle being correctly oriented and possessing the activation energy Q.

This has been further modified into theory of **Absolute reaction Rates or Transition State Theory or Activated Complex Theory.** This is based on Arrhenius concept of energy barrier and principles of quantum mechanics. For a reaction

$$A + B \ldots = C + D \ldots$$

as per the theory, first the reactants form a *activated complex* which then disintegrates into the products. The activated complex has a very short life but it exists as any other chemical species and is in equilibrium with the reactants. The specific rate of disintegration of the activated complex is a universal rate independent of the volume of reactants and the complex formed therefrom. The reaction path is often shown as in Fig. 15.2.

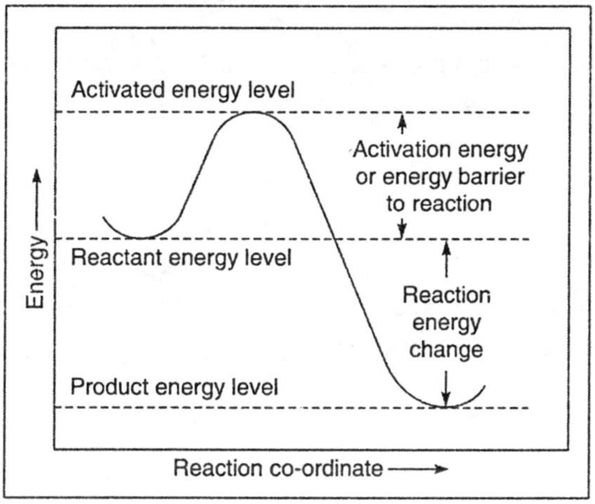

Fig. 15.2 The schematic path of the reaction showing the energy barrier.

It shows the reactants at energy level R and products at energy level P indicating that as reaction occurs R $\rightarrow$ P energy will be liberated. But before this can happen the reactants must acquire additional energy Q as activation energy and attain the activated level, *i.e.*, the activated complex of energy Q + R is formed and then the reaction occurs to produce product at P in the form

reactants $\rightarrow$ activated complex $\rightarrow$ product

The net energy release is still the same. The reactants acquire this additional energy by way of mutual collision and interaction. The reactant at times may acquire additional energy more than even Q. But if such a particle is not correctly oriented meaning thereby that the path is not due for correct collision the reaction may not occur. Hence for the reaction to take place the **minimum additional energy equivalent to activation energy and correct orientation must precede the actual collision**. The net useful equation finally has the form

$$\text{Ratio} = T\,K_0\,e^{-\Delta H/RT} \qquad \ldots(15.9)$$

This equation differs from the Arrhenius equation by an additional linear temperature term. But over a narrow temperature range the term '$T \times K_0$' can be treated as another constant and has thus the same form as that of the Arrhenius equation - which is therefore considered adequate for most purposes. The activation energy is equal to ΔH, the enthalpy change in the formation of the activated complex.

A reaction having a large negative free energy change can still occur very slowly if the activation energy is too large. The reaction is still strongly temperature dependent because of the temperature dependent exponential term.

15.3.1 Effect of Concentration on Kinetics

For a reaction $A + B \ldots = C + D \ldots$

Forward rate is $K_f \cdot C_C \cdot C_D \ldots$

and backward rate is $K_b \cdot C_C \cdot C_D \ldots$

where K_b and K_f are respectively the specific reaction rate constants of backward and forward reactions.

At equilibrium by definition itself these two rates are same and hence

$$K_f \cdot C_A \cdot C_B \ldots = K_b \cdot C_C \cdot C_D \ldots \quad \ldots(15.10)$$

dividing one by the other

$$\frac{K_b}{K_f} = \frac{C_C \cdot C_D \ldots}{C_A \cdot C_B \ldots} = K \text{ (equili. constant)} \quad \ldots(15.11)$$

For irreversible reactions the backward rate is negligibly small and hence

$$\text{Rate} = K_f \cdot C_A \cdot C_B \ldots \quad \ldots(15.12)$$

15.3.2 Effect of Temperature

From this experimental data, Arrhenius empirically showed that the rate of reaction is given as

$$\text{Rate} = A \exp^{-Q/RT} \quad \ldots(15.13)$$

where A and Q are constants. Now, the meaning of the terms have been understood as A being the frequency factor or the probability of a reactant being correctly oriented on the reaction path with energy Q. The constant Q is now known as activation energy,the minimum of which must be possessed by a particle before a reaction takes place. Taking logarithm

$$\log \text{Rate} = \log A - Q/RT$$

which shows that **log Rate vs. inverse of temperature shall be a straight line relationship.**

Reaction rates are measured at various temperatures and plotted as *rates vs. inverse of temperature.* **The –ve slope of this line is the activation energy of the process.**

15.3.3 Diffusion Laws

The **Ficks First Law** states that the flux of the diffusing species is proportional to the concentration gradient and area across which diffusion takes place or flux

$$J = -D\frac{\partial C}{\partial x} \quad \ldots(15.14)$$

where 'J' is the amount of material that diffuses per unit time in a perpendicular direction with reference plane having a unit cross-sectional area, 'C' is the concentration and x is the coordinate in the direction of diffusion and D is the coefficient of diffusion.

Ficks Second Law states that

$$\frac{\partial C}{\partial t} = D\frac{\partial^2 C}{\partial^2 x} \qquad ...(15.15)$$

where dC/dt means the rate of change of concentration.

The diffusion does not occur just because of concentration gradient. In fact thermodynamically diffusion shall occur only from where the chemical potential is high to where this is low. In other words chemical activity gradient is strictly the driving force for diffusion to occur. This is rather complicated and hence simple concentration gradient is considered sufficient to workout diffusion for general purposes.

15.4 KINETICS OF HETEROGENEOUS REACTIONS

The reaction which occurs between two or more phases are like reduction of solid iron oxide by a reducing gas,between two immiscible phases like slag and metal, leaching of solid ore by a liquid acid, distillation, and so on. In this either reactants are from different phases or the reactants and products are from different phases and as a consequence the reaction takes place only upon their contact, *i.e.*, at the interface of the two phases alone. In either case the reactants have to be transported to the interface and then interact for the reaction to occur and the products to be transported away from the interface. The process of transportation of these takes place in various ways depending upon the nature of the phase. In solid state this mass transfer is known as diffusion which is a result of movement of atoms/molecules down the concentration gradient. In a liquid phase the mass transfer can take place by either diffusion as well as by fluid flow in either lamellar or turbulent form. In lamellar way the fluid moves in the direction of flow and in turbulent by forming local eddies resulting in flow at right angles to the direction of flow.

15.4.1 Interface of Heterogeneous Systems

A solid-solid interface is merely a contact without in any way altering the structure of the two solid phases in the contact region. The story however is quite different if a fluid phase is involved.

The velocity of the fluid in contact with a solid is practically zero, although the velocity inthe bulk may be a finite value. This is because the layer in contact with the solid cannot slip over the solid surface. If it is so the flow would be a frictionless flow which is not the case. Therefore, there exist a velocity gradient from zero at the interface to whatever apparent value is there in the bulk. The fluid flow near the solid interface is thus lamellar and in the bulk turbulent with a buffer zone in between. Any reaction between solid and fluid phase therefore has to take into account this velocity gradient in the region of *velocity-boundary-layer* while considering the mass transport.

This relative stagnancy of a fluid layer at the interface, affecting mass transport also results in *concentration boundary-layer* at the interface on the fluid side. If both phases are fluid then this layer exists on either side of the interface. In case of two fluid phases in contact with each other, the interface is not a plane; it is rather wavy. There is likely to be a emulsion of both the phases, *i.e.*, dispersed phase at the interface. Although the tendency of the interface is to remain plane (as minimum energy is associated with plane surface) the eddies do not allow it to remain so.

The idealised structure of the fluid interface is shown in Fig. 15.3. Although the concentration varies continuously from the interface to the bulk, the actual contour is not readily available for mathematical treatment. Therefore, a tangent drawn at the point of interface concentration in the form of $C_I - N$ is assumed to be the concentration gradient and the $C_B - N$ is supposed to be the stagnant boundary layer for the purpose of concentration assessment.

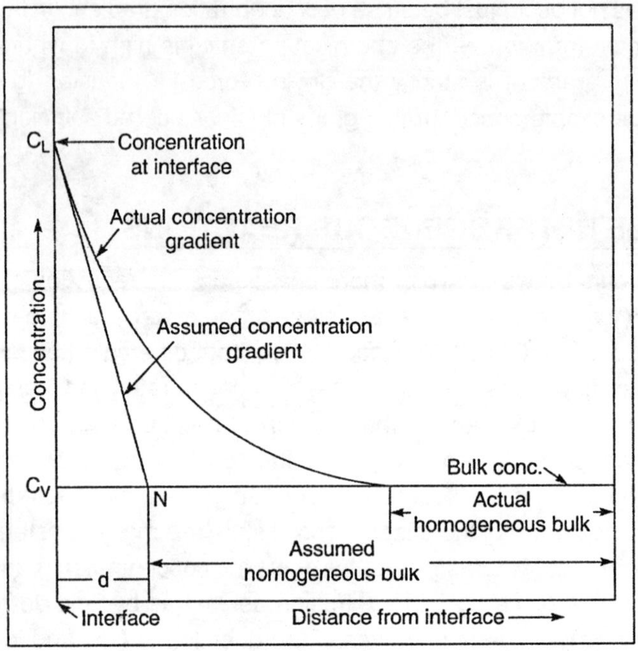

Fig. 15.3 Concentration gradient in a heterogeneous system. The assumed concentration gradient is also shown by the tangent.

The concentration gradient can now be mathematically expressed as

$$\frac{C_I - C_B}{d}$$

existing at the interface.

The diffusion can be expressed as flux

$$J = D\left(\frac{C_I - B}{d}\right)$$

The idealised structure of the interface for a two immiscible liquid phases system is shown in Fig. 15.4. It is presumed that the bulk concentration anywhere in the bulk is the same because of inherent turbulence.

If a reaction is to occur between these two phases then the reaction shall take place in the following steps :

1. Transport of reactants from the bulk across the boundary layer to the interface.
2. Adsorption of reactants into the interfacial layer.
3. Chemical reaction amongst the adsorbed species in the interfacial layer (bond exchange).

4. Desorption of products from the interfacial layer.
5. Transport of products across the boundary layer into the bulk.

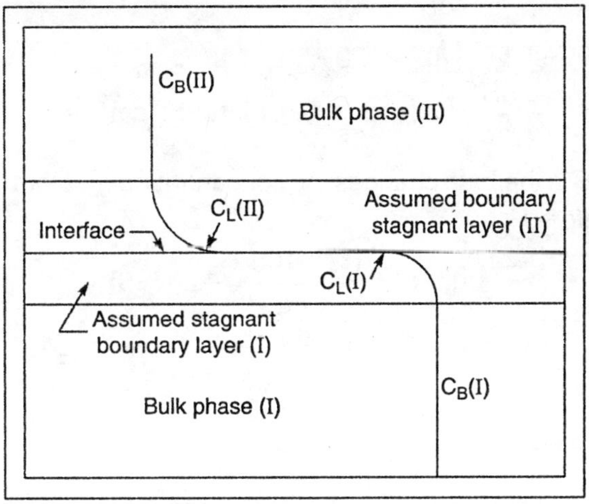

Fig. 15.4 Idealised structure of an interface of contacting fluid phases I and II.

Although physical adsorption is possible, chemisorption is more dominant in the metallurgical world chemisorption is irreversible process and it may tend to cover the whole interface and once covered fully further adsorption occurs with difficulty. The transport of reactants or products can be worked out from Fick's laws. The fraction of the interface covered by the chemisorbed reactants increases with increase in concentration of the surface active reactants in the bulk phase.

■ **Example 15.2** Find out the rate of desulphurisation of molten iron by slag in an electric arc furnace and the following conditions :

Bulk metal [% S]	= 0.06%
Interface % S	= 0.01%
Diffusion coeff. of S	$= 3 \times 10^{-4}$ cm^2/sec
Stagnant boundary layer thickness	= 0.05 cm
Average bath depth in the furnace	= 20 cm

Solution :

$$\frac{dS}{dt} = D_S \times \Delta C_S / x \text{ g/cm}^2\text{/sec}$$

where D_S is the diffusion coefficient of sulphur in metal path. x is the boundary layer thickness and ΔC_S is the difference in concentrations of bulk and equilibrium or interfacial sulphur concentration. Converting this into more convenient form in % S gives

$$-d\,\%\,S/dt = D_S \times \Delta C_S / x \left(\frac{100}{20\,\rho}\right)$$

where ρ is the density of molten iron and can be taken as 7.5 g/cc. Putting the other values in the above equation gives

$$-\%\,S/dt = 2 \times 10^{-4}\ \%/\text{sec} \quad \text{or} \quad = 0.7\ \%/\text{hr}$$

■ **Example 15.3** In the previous example find out the time required to attain 80% of the equilibrium.

Solution :

$$-\frac{dS}{dt} = D_S \times \Delta C_S / x \ g/cm^2/sec$$

converting it into more convenient form using % as

$$-\frac{dS}{dt} = D_S \times \Delta \,\%\, S/x \,(\rho/100)\ g/cm^2/sec$$

Since change in % S in the bath changes the concentration gradient responsible for diffusion of sulphur into Δ % S, therefore,

$$d[\%S] = \frac{dS \times 100}{\rho \times d} = -d[\Delta\%S]$$

or

$$dS = \frac{d[\Delta\%S] \times \rho \times d}{100}$$

putting this into the above gives

$$dt = \frac{d[\Delta\%S]}{[\Delta\%S]} \times \frac{x.d}{D_S} \quad \text{or} \quad = \frac{x.d}{D_S} \times d \ln [\Delta\% S]$$

Therefore, for a bath depth of 20 cm, boundary layer thickness of 0.05 cm and diffusion coefficient of sulphur of 3×10^{-4} and between the limits of [Δ% S] at t = 0 to be one to [Δ% S] at time 1 to be 0.20 (left over to attain equilibrium); it gives t = 158 mins.

15.5 RATE CONTROLLING STEP

Out of the above mentioned 5 consecutive steps in series for a process to occur the overall rate of the process shall be controlled or decided upon by the slowest step in this series. In general step No. 2 and step No. 4 are fast enough so that they are not considered as limiting the overall rate of reaction. Out of the remaining, two are mass transport or diffusion steps and the third one as chemical reaction step.

These various sub-steps have their individual activation energies and that step with the highest activation energy shall be the one limiting the overall rate of the reaction. If the chemical reaction as such is slow, and the rest of the steps are fast enough the overall process rate will be controlled by the rate of chemical reaction and the process is termed **as under chemical control** or it is **chemically controlled process.** On the contrary it diffusion or mass transport rate is slow and other steps are fast enough then the process is controlled by diffusion of mass transport or is said be a **diffusion or mass transport controlled process.**

This can be explained in terms of the flow of liquid in upside down position of a bottle wherein the flow rate is controlled by the constriction in the bottle and that constriction or the bottleneck in the overall flow of liquid from the bottle is controlling the flow.

The rate of diffusion or rate of chemical reaction, both are enhanced by increasing temperature of vice-versa. But the rate of diffusion is alone enhanced by stirring and it has no influence on the rate of chemical reaction.

It is very pertainment to identify the slow step in the overall kinetics of the process and therefrom work out the ways and means by which this bottle-neck step would be removed or diminished in its influence. A process under diffusion control will be accelerated by adopting stirring which would hasten diffusion. Increase in temperature would not only hasten diffusion but, if it is not controlling the rate and if the process is under chemical control, the rate as well would be improved by the temperature rise. These tricks have been fully exploited in developing a variety of steelmaking process techniques for improving kinetics of steelmaking reactions and thereby the productivity of steelmaking shops.

Activation energy values are often taken as an indication of the type of control possibly existing in the process. If the process is complex and if it is not possible to simplify it for laboratory studies for estimation of activation energy values, the apparent rates vs. inverse of temperature are plotted and from then on,the apparent activation energy is estimated. Even this gives approximate indication of the possible control existing in the process. Low values of activation are often taken as an indication of diffusion controlled processes and high values for chemically controlled processes. The intermediate values can however be misleading and should be interpreted carefully.

15.5.1 Mechanism of Solid-Fluid Reactions

The mechanism of solid-fluid interfacial reaction is shown in Fig. 15.5. In such reactions. If the product of the reaction is fluid it gets assimilated in the bulk fluid and by transport through stagnant fluid layer the reaction may be controlled by such transport of reactions or products or by chemical reaction as such. Gassification of carbon by CO_2 as

$$<C> + \{CO_2\} = 2\{CO\} \quad ...(15.16)$$

is a typical example of this type and the reaction has been found to be controlled by chemical reaction step.

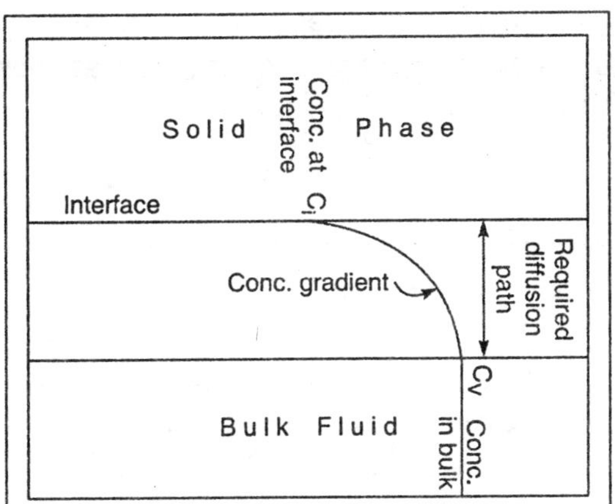

Fig. 15.5 The idealised concentration gradient in a solid-fluid system when the product of the reaction does not accumulate at the interface.

The other type is where product is a solid condensed phase and in that case the situation after a certain amount of reaction has occurred is shown in Fig. 15.6.

The reaction results in accumulation of product layer which separates the main interacting phases. Further reaction can occur only by diffusion of reactants across this product layer, in addition to the usual diffusion of reactants across the stagnant fluid layer. In this case the

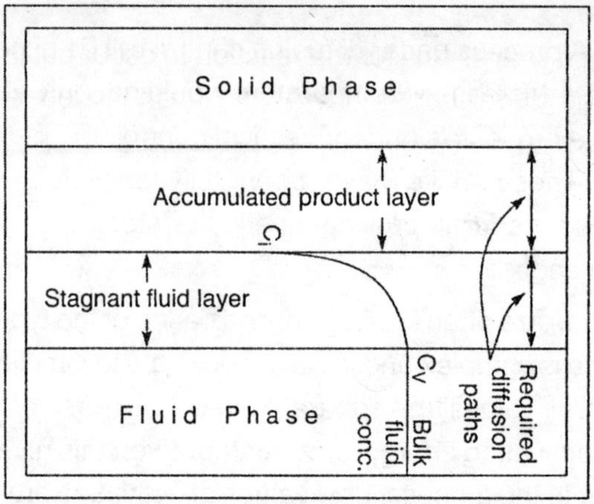

Fig. 15.6 The idealised concentration gradient in a solid-fluid system when the product of the reaction does accumulate at the interface.

accumulation may result in contraction of the area of original interface depending upon how the accumulation occurs. The shape of the interfacial area however remains the same, as like the original. Such processes where the shape of the original reaction interface remains the same but actual area is altered are known as **topochemical processes.** The rate of such reaction will decreases continuously with the progress of the reaction because of increasing accumulation of the product at the interface. The reduction of Fe_2O_3/Fe_2O_4 by H_2 or CO gas is a typical case of this type. It can be shown that if **$[1-(1-F)^{1/3}]$ vs. time** is a straight line then it is a process controlled by chemical reaction and if $\left[1-\frac{2}{3}F-(1-F)^{2/3}\right]$ vs. time is a straight line then it is under diffusion control; where F is the fractional reaction occurred at time t. This is true when the reaction interface is well defined and remains so during the progress of the precess. However, if the solid particle is porous then the kinetics will be affected by the **tortuosity factor** (interlinking of pores) and thepore volume.

15.5.2 Nucleation and Growth

If reactions or processes result in the formation of a new phase then the nucleation and growth of that phase has to be taken into account while evaluating their kinetics. The nuclei are too small and require volume free energy and interfacial free energies to be supplied from the process for its formation. It is not easy to obtain so much energy and hence it is not easy to nucleate a new phase homogeneously. But if part of this energy requirement of nucleation is met by providing readymade interface the nucleation becomes easier. The necessity of homogeneous nucleation is a barrier and can prevent homogeneous reaction from taking place at all. The typical case is of carbon-oxygen reaction in steelmaking. If an interface is made available for this then the reaction rates have been observed to be reasonably large. The rate of reaction of oxidation of carbon during steelmaking as

$$[C] + [O] = \{CO\} \quad \text{...(15.16)}$$

is very slow when iron ore is added for its oxidation but when gaseous oxygen is supplied for its oxidation it is much faster. This has been attributed to the fact that when gaseous oxygen is supplied the gas-metal interface is readily available for the reaction to occur but this is not the case when iron oxide is added and hence the sluggishness of this reaction in the latter case.

15.5.3 Reaction Rates and Heat Supply

A process may be endothermic such that unless adequate heat supply is available the reaction may not take place for want of heat. The case of carbon oxidation as in Eq. (15.16) in open hearth process of steelmaking where oxygen is made available from Fe_2O_3, making the overall process occur as

$$<Fe_2O_3> + 3[C] = 2[Fe] + 3\{CO\} \qquad ...(15.17)$$

which is a net endothermic and hence simultaneous heat supply has to be assured then only it may be under diffusion or chemical control. But if heat flow itself is lacking for the available diffusion rate or chemical reaction rate, it is the heat rate that will then be controlling the overall rate of this reaction. This rate of heat transfer depends on temperature difference between the source of heat and the reaction interface, physical geometry and thermal constraints of such a reaction. But the rate of such a reaction like oxidation of carbon can be enhanced by converting this net endothermic reaction into a net exothermic reaction if oxygen gas is supplied in place of iron oxide for oxidation when the net reaction that takes place is as given in Eq. (15.16) which is net exothermic in nature.

15.6 CONCLUDING REMARKS

The study of rates of processes is a must for understanding the time required for carrying out such processes in actual practice. This has a direct bearing on the process economy. The understanding of the rate controlling steps can and has actually resulted in improving the process economy by reducing the time required for carrying out these processes in practice. It also indicates the practical limitations of improving the process economy by improving the productivity.

□□□

BIBLIOGRAPHY

Books Recommended for Additional Reading

1. Physical Chemistry of Metals, by L.S. Darken and R.W. Gurry, McGraw Hill, 1953.
2. Chemical and Metallurgical Thermodynamics by M.L. Kapoor, Vol. I and II, Nemchand and Bros., Roorkee, 1979.
3. Introduction to Metallurgical Thermodynamics by D.R. Gaskell, McGraw Hill, Tokyo, 1973.
4. Thermochemistry of Steelmaking by J.F. Elliot, M. Gleiser and Ramkrishna Vol. II, Addison Wessley, London, 1963.
5. Problems in Metallurgical Thermodynamics and Kinetics by G.S. Upadhyaya and R.K. Dube, Pergamon Oxford, 1977.
6. Classical Thermodynamics by A.B. Pippard – 1963.
7. Physical Chemistry of Iron and Steel Making by R.G. Ward, Edward Arnold, 1962.
8. Physical Chemistry of Iron and Steel Manufacture, by G. Bodsworth, Longman's, 1963.
9. Metallurgical Thermochemistry by O. Kubaschewski, E.L. Evans and C.B. Alcock, Pergamon – 1969.
10. Physical Chemistry of Melts in Metallurgy by F.D. Richardson, Vol. I and II Academic Press, Londo, New York, 1974.

□□□

SYMBOLS

Activation energy	Q
Activity - Raoultian	a
Henrian	h
Activity coefficient Raoultian	γ
Henrian	f
Chemical potential	μ
Concentration	C
Diffusion coefficient	D
Entropy	S
Electrochemical valency	n
Electro-motive-force	$\mathcal{E}$
Equilibrium constant	K
Entropy	S
Excess property	xs
Farraday's constant	$\mathcal{F}$
Finite change in extensive property	Δ
Fugacity	f
Gas constant	R
General extensive property	Z
Gibb's free energy	G
Gram	g
Heat capacity	C
Heat exchanged	q
Helmholtz free energy	A
Internal energy	U
Interaction parameter	e_i^i
Meter	m
Mixing function	Z^M
Partial property	$\bar{Z}$
Specific heat constant volume	C_V
constant pressure	C_P
Standard state with superscript	o
Specific rate constant	k
Temperature degrees absolute	°K
degrees Celcius	°C
Volume	V
Work done mechanical	w
any other	w′

INDICATORS

Composition in weight percentages	wt%
Gas phase	{ }
Liquid oxide phase	( )
Molten metallic phase	[ ]
Solid phase	< >

ABBREVIATIONS

atmos	atmospheric pressure
cal	calories
Kcal	kilo-calories
cc	cubic centimeters
hr	time in hours
ln	natural logarithm
log	logarithm to the base ten
min	time in minutes
sec	time in seconds

CONVERSION FACTORS

Farraday's constant, F	96,500 coulombs/g equivalent
	23,061 calories/volt/g equivalent
Gas constant R	1.987 cal/deg/mole
	8.314×10^7 ergs/deg/mole
	82.07 cc atmos/deg/mole
	0.08207 litre atmos/deg/mol
ln	2.303 $\log_{10}$
One atmosphere pressure	760 mm Hg
	760 torr (1 mm = 1 torr)
	1033 g/cm^2

INDEX

G

H

I

K

L

M

S

T

□□□